# THE PROCESS OF SCIENCE

# SCIENCE AND PHILOSOPHY

This series has been established as a forum for contemporary analysis of philosophical problems which arise in connection with the construction of theories in the physical and the biological sciences. Contributions will not place particular emphasis on any one school of philosophical thought. However, they will reflect the belief that the philosophy of science must be firmly rooted in an examination of actual scientific practice. Thus, the volumes in this series will include or depend significantly upon an analysis of the history of science, recent or past. The Editors welcome contributions from scientists as well as from philosophers and historians of science.

NANCY J. NERSESSIAN

*Editor*

# The Process of Science
*Contemporary Philosophical Approaches to*
*Understanding Scientific Practice*

1987 **MARTINUS NIJHOFF PUBLISHERS**
a member of the KLUWER ACADEMIC PUBLISHERS GROUP
DORDRECHT / BOSTON / LANCASTER

**Distributors**

*for the United States and Canada*: Kluwer Academic Publishers, P.O. Box 358, Accord Station, Hingham, MA 02018-0358, USA
*for the UK and Ireland*: Kluwer Academic Publishers, MTP Press Limited, Falcon House, Queen Square, Lancaster LA1 1RN, UK
*for all other countries*: Kluwer Academic Publishers Group, Distribution Center, P.O. Box 322, 3300 AH Dordrecht, The Netherlands


**Library of Congress Cataloging in Publication Data**

```
The Process of science.

   (Science and philosophy)
   Includes index.
   1. Science--Philosophy.  2. Science--Methodology.
I. Nersessian, Nancy J.  II. Series.
Q175.3.P785  1987        501              86-23666
ISBN 90-247-3425-8
```


ISBN 90-247-3425-8

PRINTED IN THE NETHERLANDS

# Table of contents

VI

# Preface

For some time now the philosophy of science has been undergoing a major transformation. It began when the 'received view' of scientific knowledge – that developed by logical positivists and their intellectual descendants – was challenged as bearing little resemblance to and having little relevance for the understanding of real science. Subsequently, an overwhelming amount of criticism has been added. One would be hard-pressed to find anyone who would support the 'received view' today. Yet, in the search for a new analysis of scientific knowledge, this view continues to exert influence over the tenor of much of present-day philosophy of science; in particular, over its problems and its methods of analysis.

There has, however, emerged an area within the discipline – called by some the 'new philosophy of science' – that has been engaged in transforming the problems and methods of philosophy of science. While there is far from a consensus of beliefs in this area, most of the following contentions would be affirmed by those working in it:
- that science is an open-ended, on-going activity, whose character has changed significantly during its history
- that science is not a monolithic enterprise
- that good science can lead to false theories
- that science has its roots in everyday circumstances, needs, methods, concepts, etc. of human beings
- that through examination of science we will learn how to understand such notions as 'observation', 'theory', 'meaning', 'reference', 'explanation', 'progress' and 'rationality'
- that the boundary between a 'scientific' question and a 'philosophical' question, especially as regards foundational problems, is often blurry
- that the function of the philosophy of science is descriptive, critical, and normative
- that the philosophy of science is not self-sufficient, insofar as the methods of philosophy need to be supplemented by concepts, principles, theories, etc., drawn from other disciplines; though it remains to be determined just what can be used and how.

These beliefs have as support extensive analyses of 'case studies' of past and present scientific practices. Such examinations have shown the importance of 'case study' analysis to go far beyond that of simply providing a means of critiquing philosophical views. That is, study of the processes by which theories and concepts are formulated and accepted or rejected is prerequisite to philosophical understanding. Rather than creating sweeping theories, the 'new philosophy of science' is attempting to deal with the existing complexities of actual scientific practices. But *how* are philosophers to do this? By what methods can philosophers legitimately inquire about the processes of scientific inquiry and its products? Resolving this question is of fundamental importance to the future of philosophy of science.

When the *Science and Philosophy* series was established, it was decided that one of the early volumes should contain a collection of essays in which philosophers were offered an opportunity to reflect upon the methods that they are using and believe should be used to gain the fullest and most accurate understanding of the nature of scientific knowledge and inquiry. Although questions of method in the philosophy of science form the common thread running through these essays, the views expressed and the topics discussed are quite varied. The result is a collection which, while not exhaustive of contemporary views in the area, gives witness to both the diversity of viewpoint and the commonality of concerns of the 'new philosophy of science'. As such it provides a valuable sourcebook of contemporary methodological practices.

I thank the contributors for undertaking a difficult task and for producing a stimulating and provocative volume.

I would like, also, to take this opportunity to thank Martinus Nijhoff Publishers and, in particular, Alexander Schimmelpenninck, for enabling me to establish this series.

Finally, I appreciate the hospitality of the Center for the Philosophy of Science, University of Pittsburgh, during the final stages of the preparation of this volume.

Pittsburgh, May, 1986                                          Nancy J. Nersessian

# List of contributors

Joseph Agassi, Tel Aviv University & York University; Address York University: Dept. of Philosophy, 4700 Keele Street, Downsview, Ontario, M3J IP3, Canada

William K. Berkson, 1618 Waters Edge Lane, Reston, VA 22090, U.S.A.

Harold I. Brown, Northern Illinois University, Dept. of Philosophy, De Kalb, IL 60115, U.S.A.

Richard M. Burian, Virginia Polytechnic Institute and State University, Dept. of Philosophy, Blacksburg, VA 24061, U.S.A.

Bas C. van Fraassen, Princeton University, Dept. of Philosophy, Princeton, NJ 08540, U.S.A.

Ronald N. Giere, Indiana University, History & Philosophy of Science, 130 Goodbody Hall, Bloomington, IN 47405, U.S.A.

Marjorie Grene, 206 Ridgedale Road, Ithaca, NY 14850, U.S.A.

Sandra Harding, University of Delaware, Dept. of Philosophy, Newark, DE 19716, U.S.A.

Nancy Nersessian, University of Pittsburgh, Dept. of Philosophy, 817 Cathedral of Learning, Pittsburgh, PA 15260, U.S.A.

Thomas Nickles, University of Nevada-Reno, Dept. of Philosophy, Reno, NV 89557, U.S.A.

Dudley Shapere, Wake Forest University, Dept. of Philosophy and History of Science, Drawer 7229 – Reynolda Station, Winston-Salem, NC 27109, U.S.A.

X

Elliott Sober, University of Wisconsin, Dept. of Philosophy, 5185 White Hall, Madison, WI 53706, U.S.A.

# Notes on the contributors

*Joseph Agassi* holds joint appointments as Professor of Philosophy at York University, Ontario, Canada, M3J 1P3, and Tel Aviv University, Tel Aviv, Israel. He is author of numerous articles and books, among which are *Science in Flux* and *Science and Society*.

*William Berkson* did his Ph.D. in the History and Philosophy of Science at the London School of Economics under Sir Karl Popper. He is author of *Fields of Force: The Development of a World View from Faraday to Einstein* and (with J. Wettersten) *Learning from Error: Karl Popper's Psychology of Learning*. He is now an independent writer (1618 Waters Edge Lane, Reston, VA 22090, USA). He is presently writing a book on personal decision-making in the face of opportunity and risk.

*Harold Brown* is Professor of Philosophy, Northern Illinois University, De Kalb, IL 60115, USA. He has published in many areas of philosophy of science, and is author of *Preception, Theory and Commitment: The New Philosophy of Science* and *Observation and Objectivity*.

*Richard Burian* is Head of the Department of Philosophy, Virginia Polytechnic Institute and State University, Blacksburg, VA 24061, USA. His publications concentrate on conceptual change in science and philosophy of biology. With Robert Brandon he co-edited *Genes, Organisms, and Populations*. He is currently at work, with geneticist Doris Zallen, on *A Conceptual History of the Gene* and, with Richard Burkhardt, Jr. and Richard Lewontin, is co-editor of a monograph series in history and philosophy of biology for Oxford University Press.

*Bas C. van Fraassen* is Professor of Philosophy, Princeton University, Princeton, NJ 08540, USA. His publications include *An Introduction to Space and Time* and *Formal Semantics and Logic*. His most recent book, *The Scientific Image,* was co-winner of the 1981 Matchette Prize for outstanding scientific merit.

*Ronald Giere* is Professor of History and Philosophy of Science, Indiana University,

Bloomington, IN 47401, USA. He has been an Associate Research Scientist at the Courant Institute of Mathematical Science, New York University and a Senior Research Fellow at the Center for Philosophy of Science, University of Pittsburgh. He has published widely on induction, probability, causality, and statistics, and on the relations between history and philosophy of science. He is author of *Understanding Scientific Reasoning,* and co-editor of *Foundations of Scientific Method: The Nineteenth Century,* and of *PSA 1980.* His current project is *A Cognitive Theory of Science,* a study of representation, judgment, and tradition in modern science.

*Marjorie Grene* is Professor Emeritus, the University of California, Davis, CA, 95616, USA. She has written extensively in the history of philosophy, epistemology, philosophy of science, and philosophy of biology. Her books include *Approaches to the Philosophy of Biology* and *The Understanding of Nature: Essays in the Philosophy of Biology.* Her most recent edited collection (with Debra Nails) is *Spinoza and the Sciences.*

*Sandra Harding* is Associate Professor of Philosophy, University of Delaware, Newark, DE 19716, USA, where she also has a joint appointment in Sociology and is Director of Women's Studies. She is editor of *Can Theories Be Refuted? Essays on the Duhem – Quine Thesis* and of *Discovering Reality: Feminist Perspectives on Epistemology, Metaphysics, Methodology, and Philosophy of Science* (with Merill B. Hintikka). Her forthcoming book is *The Science Question in Feminism.*

*Nancy Nersessian* is presently Senior Research Fellow at the Center for Philosophy of Science, University of Pittsburgh, Pittsburgh, PA 15260, USA. She has been a Fulbright Research Scholar at the University of Leiden and a Fellow of the Netherlands Institute for Advanced Study. Her publications include *Faraday to Einstein: Constructing Meaning in Scientific Theories.* She is in the process of editing five volumes of the works of H.A. Lorentz on electromagnetism.

*Thomas Nickles* is Professor of Philosophy, University of Nevada, Reno, NV 89557, USA. He has published many papers on explanation, intertheoretic reduction, and scientific discovery and problem solving. He has edited *Scientific Discovery, Logic, and Rationality* and *Scientific Discovery: Case Studies.*

*Dudley Shapere* is Z. Smith Reynolds Professor of the History and Philosophy of Science, Wake Forest University, Winston-Salem, NC 27109, USA. He received his B.A., M.A., and Ph.D. from Harvard University, and has taught at the Ohio State University, University of Chicago, University of Illinois, and University of Maryland. His most recent book is *Reason and the Search for Knowledge.*

*Elliott Sober* is Professor of Philosophy, University of Wisconsin, Madison, WI 53706, USA. He held a Guggenheim Fellowship during 1980–1, during which time he was a Fellow in Richard Lewontin's population genetics laboratory at the Museum of Comparative Zoology at Harvard University. He is the author of *The Nature of Selection* and editor of *Conceptual Issues in Evolutionary Biology*.

# Method in the Philosophy of Science and Epistemology
## How to Inquire about Inquiry and Knowledge

DUDLEY SHAPERE

*Wake Forest University*


# I

To those attempting to understand the knowledge-seeking and knowledge-acquiring enterprise – to understand how we should go about trying to get knowledge, and what it is that we have if we get it – one of the major lessons of science in the twentieth century is this:

> The results of scientific investigation could not have been anticipated by common sense, by the suggestions of everyday experience, or by pure reason.

I will call this the *Principle of Rejection of Anticipations of Nature*. The name derives from the writings of Francis Bacon, who long ago warned against anticipations of nature.[1] Bacon, however, had far more primitive sorts of anticipations in mind than those brought to the fore by twentieth-century science, and in any case he had a very imperfect understanding of the significance of the principle. In his conception, the principle was a means of ruling out certain types of ideas, rather than, as with the modern version of the principle, of opening the door to unforeseen possibilities. It must be added, too, that Bacon's writings are filled with anticipations of how nature and its investigation must or cannot be.

The full significance of the principle is a product of scientific inquiry, and especially of the results of that inquiry in this century; it is not an *a priori* stricture, laid down, for example, by the nature of inquiry or scientific method itself. It might be obviated by future inquiry. Perhaps we will someday discover a room (call it a Cartesian room) in which whatever occurs to the occupant turns out to be correct, so that we come to believe that had we found that room earlier, we could have found all truth by its use. Or it might turn out that there is only one mathematically possible physical theory, and that purely mathematical inquiry, coupled with relatively unsophisticated empirical considerations, might have led to its discovery. The second possibility perhaps appears more realistic than the first, but only because there are some very tentative indications that there may be only one possible form of mathematical theory of elementary processes – but not that we might have found

*Nancy J. Nersessian (editor), The Process of Science. ISBN 90-247-3425-8.*
© *1987, Martinus Nijhoff Publishers (Kluwer Academic Publishers), Dordrecht. Printed in the Netherlands.*

that theory by any means other than the arduous investigations by which we will in fact have been led to it should that occur. In the case of neither possibility, or of any other imaginable one, is there the slightest reason to suppose the Principle of the Rejection of Anticipations of Nature to be false.

The evidence in its favor is indeed overwhelming; even a brief survey of just some of the more familiar such evidence can only lead to its acceptance. Conflicts between electromagnetic theory and mechanics were resolved by a deeper understanding of the nature of spatial and temporal measurement, an understanding that resulted in a fusion of the concepts of space and time into that of space-time, a fusion that utterly contradicted the preconceptions of earlier thought. General relativity brought space-time into causal connection with matter, a connection whose very conceivability had been denied by Newton. Based on the concept of intrinsic curvature of a space introduced by Gauss and generalized by Riemann and his successors to a broadened and deepened conception of what it is for something to be a space, general relativity in its cosmological guise made possible the idea of the expansion of space – an expansion not *into* some containing space, but *of* the space itself. Given the dependence of such everyday ideas as expansion and curvature on the idea of a containing space in which expansion must take place and with respect to which curvature must be determined, such an idea was inconceivable, even self-contradictory, beforehand; both Newton and Leibniz, only remotely approaching a conception of the possibility of intrinsic and varying properties of space in connection with their criticisms of Descartes, dismissed it as absurd. The concept of explanation itself underwent a series of transformations following seventeenth- and eighteenth-century arguments that that 'very concept' required genuine explanation to be mechanistic (i.e., in terms of matter and motion) and deterministic (i.e., in terms such that complete and precise prediction and retrodiction should be possible). Following a long history of reinterpretations and modifications of this view, quantum mechanics and its extensions in quantum field theory and its relatives altered it profoundly, ultimately bringing the concept of symmetry and its breaking to centrality in explanation. Along the way, ideas of symmetries which had been so taken for granted that they had not even been formulated were brought out and given explicit statement; and sometimes, as in the case of parity, the explicit formulation exposed the superficiality of the previously inexplicit concepts. Once formulated and connected with fundamental physics, those concepts were tested and, again as in the case of parity, sometimes rejected, at least as universal truths. The universe has been found to contain types of entities whose very possibility was not and in many cases could not have been conceived by earlier thinkers. And although some earlier writers had spoken of the universe as 'infinite,' such references failed to anticipate the vastness of the universe of modern science. The sizes of galaxies and their associations, and of interstellar and intergalactic space and the spaces between clusters of galaxies, and, even more importantly, the ways in which the global and the local structure of the universe are related, are far beyond anything those earlier writers would have supposed in their wildest imaginings. And indeed, the very question of the finitude or infinitude of the universe has undergone a radical reformulation through a deep revision of the concept of infinitude, one which, among other things, distinguishes infinitude from unboundedness. The question of the dimensionality of space (and space-time) was given physical significance by nineteenth-century developments in mathematics and by the special theory of relativity; but attempts to unify the fundamental forces of nature have led from the primitive discussions by Einstein and Kaluza to present discussions in which serious consideration is given to such entities as 26-dimensional spaces, in which the mathematical concept of compactification is given not only physical significance but even dynamical interpretation. The elementary particles of this universe have properties and exhibit behavior which could not have been thought of before the introduction and development of quantum mechanics. In connection with that theory and its field-theoretic extensions, quantum theories have begun to revise our very concept of existence itself, as quantum and relativistic cosmologies are in the process of altering our understanding of the term 'universe'.

Nor is such evidence limited to the physical sciences. Who could have anticipated the complexity of the processes of life – of heredity and development, or of the nervous and immune systems? Assumptions about what the distinctions and relations between the living and the non-living are, or even must be, have been drastically revised in the process of our learning about those processes and systems. Views prevalent among geologists past the middle of the present century, according to which the history of the earth's surface is produced by vertical forces like erosion and isostasy, and perhaps contraction or expansion of the earth, have been overturned in favor of the 'ridiculous' and 'unscientific' view that the primary forces involved are horizontal ones.

But if so many of our contemporary scientific beliefs could not have been anticipated by common sense, the suggestions of everyday experience, or pure reason, then how *have* we managed to think of them and come to adopt them? Despite its undeniable importance for our understanding of some aspects of the scientific process, the Principle of Rejection of Anticipations of Nature is essentially negative, telling us that we cannot take for granted even those of our beliefs that seem most obvious. It does not tell us how we have managed to depart from our everyday and common sense beliefs and from the presumptive dictates of reason and ultimately come to our present scientific beliefs. Still less does it tell us whether (and if so, why and in what sense) that process is a rational one, whether we are justified in holding the beliefs at which we have arrived, or whether those beliefs can be taken as true of nature. That understanding is gained through a second great lesson of modern science, one which furnishes profound insight into not only the knowledge-seeking but also the knowledge-acquiring aspect of the scientific enterprise.

This positive lesson of modern science stems from the fact that the sorts of considerations that have led us to alter our beliefs about nature, at least when those considerations are ones we call 'rational' or 'based on evidence', have themselves been scientific ones. For twentieth-century science, even more than its predecessors, has shown the possibility of formulating our beliefs in ways that make it possible to subject them to scientific scrutiny, and thereby to see how we might modify or reject and replace them if necessary. Bringing them thus to scientific investigation involves showing that and how the beliefs in question are relevant to at least some well-founded scientific beliefs and practices, and how those beliefs and practices are relevant to them. In case after case, especially in twentieth-century science, it has proved possible to show such relevance; and time and time again it has turned out that when such relevance has been shown, the previously tacit or cherished assumptions in question have had to be modified or rejected. It is thus through this incorporation of beliefs into the scientific process that it has become possible to modify or reject so many beliefs which had previously seemed unassailable, and to have arrived at so many beliefs of modern science which could not otherwise have been anticipated – and, as we shall see in what follows, to have done these things for good reasons. This record of achievement provides the second major lesson of modern science, a lesson which can be formulated as a principle:

> Every aspect of our beliefs ought, wherever possible, to be formulated, and to be brought into relation to well-founded beliefs, in such a way that it will be possible to test that aspect.

In short, this second lesson of modern science tells us to *internalize* all aspects of our beliefs into the scientific process. For that reason, I will call it the *Principle of Scientific Internalization,* or, more briefly, the *Principle of Internalization.* It is a *normative* principle; and its value, its necessity, as a policy, a guiding principle, of science is something that has itself been learned through the scientific process, through a record of achievement that led to its adoption.

In connection with the two lessons or principles, it is important to realize that radical changes in the fabric of science have not been restricted to alterations of our substantive beliefs about how things are. They have also extended to the methods and rules of reasoning by which we arrive at those beliefs, and the aims we have in seeking them. I have already mentioned some ways in which what was thought to be necessary to explanation has changed: the abandonment of mechanistic and deterministic constraints on what could count as a genuine explanation, and the ascendancy of the idea of symmetry and symmetry-breaking to being the leading idea involved in ultimate physical explanations. In other writings I have examined some other aspects of this question: the rise of the empiricist requirement that every detail of our experience should be subject to precise – and quantitative – explanation; the replacement of a perfectionist by a compositionalist approach to explanation of material substances; and relaxations of the idea that explanation must consist in rigorous deduction.[2] The import of these examples is by no means restricted to the concept of explanation. Kepler's approach was empiricist not merely in its demand for explanation of the details of experience, but also in its demand for close observation in order to obtain those details. The transition from perfectionist to compositionalist theories of explanation involved a shifting of the very goals of the study of matter, and was associated with a shift in the ways material substances are to be classified, and with the language in terms of which they are to be described.[3]

A host of other convincing examples exist. In other work, I have tried to show three major ways in which the concept of observation has come to be integrated into the fabric of scientific belief. First, the status of observation as the primary means of testing beliefs and obtaining knowledge was something that had to be established, in competition with other claimed sources of knowledge. Second, the interpretation of what counts as observational has come more and more to depend on the content of the very scientific beliefs to which it itself has led. (Correspondingly, as the evidential role of observation became more and more emphasized, the original connections between perception and observation were increasingly loosened.) And third, the exact ways in which observation plays its evidential roles – of confirmation and disconfirmation, verification and falsification – have been shaped by the content of beliefs about the world whose warrants have themselves come from observation.[4] It is truly *all* aspects of science, not only what are considered its substantive beliefs about nature, but also its methods and aims, that are subject to change in ways that have continued to surprise us. The problems we face in our inquiries about nature, and the methods with which we attempt to deal with those problems, co-evolve with our beliefs about nature. We have not been able to anticipate how to learn and to think about nature, and what questions to ask, any more than we have been able to anticipate what we would learn. The methods we employ lead us to new beliefs, which in turn lead us to modify those very methods, sometimes replacing them with new ones. The result has been a growing integration of method with belief. A cycle of mutual adjustment of beliefs and methods has thus become a characteristic feature of the scientific enterprise; it is a process of *internalization* of methodology into the rest of science.[5]

Nor is the process of internalization limited to the gradual integration of methodology into the rest of science. What counts as a reason, too, has become a function of scientific belief – belief which itself has been attained by a process of reasoning. As I have tried to show in a number of writings,[6] the distinction between the scientifically relevant and the scientifically irrelevant is one that has had to evolve, leading gradually to a distinction between considerations which are 'internal' to science and ones which are 'external' to it. (Attempts to draw an unalterable 'line of demarcation' between the scientific and the non-scientific are as much anticipations of nature as any of the other examples discussed above.) And I have argued that those considerations which have proved successful and free of specific and compelling doubt are those which become internal to the scientific process and are the considerations upon which further problems and hypotheses are constructed, methods established, patterns of explanation conceived, and goals delineated.[7] (Criteria of 'success' and 'doubt' are also developed as functions of the scientific enterprise.[8]) And, finally, I have argued that such considerations count as *reasons* in science.[9]

The problems that count as scientific, too, alter with the development of science. Questions once considered scientifically legitimate – how to bring matter to perfection; the final causes of things – have been abandoned in the light of new successful and doubt-free beliefs. Questions earlier dismissed as being improper subjects for scientific investigation – the origin of life, the origin of the stars, the chemical composition of the stars, the origin of the chemical elements, the origin of the universe – have been found to be related to successful and doubt-free beliefs in such ways that they have become scientifically tractable problems. They have, in other words, become internalized into the scientific process.

Even such lofty concepts as 'truth' and 'existence' have been refashioned, or are being refashioned, in the light of what we have come to believe in science. As with all our concepts, even the most categorical, these are products of our human ('middle-sized') experience; and such concepts are, as we have had to learn, not necessarily appropriate to the understanding of nature. They have become (or rather, are being) attuned to the knowledge-seeking enterprise in the light of the results of that enterprise, as they in turn have contributed to altering the structure of scientific belief.[10]

The goal of internalization has been achieved to a very considerable extent in modern science, even if not yet completely. (The possibility of achieving internalization is itself a contingent matter, not establishable *a priori*.) That achievement serves an important function in the knowledge-seeking enterprise. By bringing methodology, reasoning-patterns, goals and so forth into intimate connection with more overtly substantive claims, it makes it possible to subject all the aspects of science to test. But the need to bring all aspects of scientific inquiry, even the methods, goals, reasoning-patterns, and standards of explanation, to test, is precisely what I described as the second lesson of modern science. That is why I have called that lesson the *Principle of Internalization*.

## II

These two lessons of modern science, the Principle of Rejection of Anticipations of Nature and the Principle of Internalization, have deep implications for the philosophical study of science and its conclusions. But, as I shall try to show, the implications are far from having been fully grasped; the lessons have not been learned.

The most important manifestation of the failure of philosophers of science to grasp these lessons is found in the continued prevalence of what may be called the *Levels Approach* to the philosophical interpretation of science. This approach rests on the following assumption: that *understanding of what science is (or does) must be independent of the content of any specific scientific beliefs.* Particular scientific beliefs come and go; but what makes them scientific must be some common feature which is shared by all possible scientific practices or beliefs. (It is thus an essentialist approach to science.) Therefore the analysis of what science is and does, in the most general sense, has to do with a 'level' of ideas, methods, rules, or whatever which governs the procedures of science itself, and is in that sense 'above' and independent of the scientific process itself.[11]

Levels approaches in general are faced with a number of difficulties, each serious enough by itself but compelling when taken together. Even apart from the question as to whether they really are independent of the scientific testing process – a question which has in the past been answered negatively for so many candidates for upper-level status – there is the problem of how the allegedly upper-level concepts are to be justified. To hold that they are waranted by their scientific fruits is to abandon their independence of the scientific process itself. But on the other hand, attempts at *a priori* or transcendental deductions of them have failed uniformly and completely, and in any case such deductions are fundamentally anticipations of nature, and certainly anticipations of what methods will enable us to find out about nature.

In order to bring out more fully what is involved in the levels type of approach, and also the difficulties which such approaches face, let us examine some of its more common versions. One type of example consists of theories according to which the very possibility of scientific inquiry requires the presupposition of certain factual assertions, assertions which, because they are presupposed by the scientific enterprise, could never be overthrown by such inquiry. This type of levels view is no longer tenable. The assertions that have been proposed as necessary presuppositions of the scientific enterprise have been universally shown either to be empty, as in the case of the Principle of the Uniformity of Nature, or else subject after all to the testing procedures of science. Indeed, all too often, as in the case of the assertion of determinism, they have been rejected as false. That fact by itself has made many rightly suspicious of this sort of view.

A second type of levels view is one according to which there are certain concepts, including 'theory', 'evidence', 'observation', 'confirmation', 'explanation', which are used in talking about science, as opposed to concepts like 'force', 'space', 'atom', 'bond', or 'ribosome', which are employed in science itself. The former type of concepts are called 'metascientific', the latter, 'scientific'. And according to this type of view, *anything* – past, present, future, or possible – that qualifies as a scientific theory, or a scientific explanation, etc., must satisfy the definitions of the metascientific terms. The totality of metascientific concepts, properly analyzed, thus define what science is, forever.

This 'metascientific' approach was the one adopted by logical empiricists. It was their further contention that the analysis of metascientific concepts should be in terms of formal logic. Thus a scientific theory was seen as an interpreted axiomatic system, the 'rules of interpretation' themselves being analyzable (at least in the most consistent and intuitively appealing version of this view) in purely logical terms. Or again, scientific explanation was 'explicated' as follows: what is explained in science is deduced logically from premises which include a 'lawlike' statement, the analysis of 'lawlikeness' itself being made in purely logical terms. But in the case of every attempted analysis of metascientific concepts, the purely logical analysis either gave a highly distorted picture of science, or else was so empty of content as to provide little illumination. Any required supplement to make the analyses more adequate (for example, in the case of explanation, any supplement distinguishing lawlike from non-lawlike statements, or any modifications to show why so many explanations are not deductive and cannot be put into deductive form) could only come from the content of science itself. Even to the extent that the deductive theory of explanation could be considered applicable to science, its account of explanation was so empty and superficial that the really important issues in scientific explanation, the ones where real illumination of scientific reasoning might be obtained (e.g., the reasoning involved in the assumption of, and ultimate abandonment of, mechanistic and deterministic requirements on explanation; the reasoning leading to the ascendancy of symmetry and symmetry-breaking forms of explanation) were left completely untouched. Finally, if the Principle of Rejection of Anticipations of Nature and the Principle of Internalization are accepted, even logic, while it can be used as a basis for building our ideas in and of science, must itself be viewed as subject in principle to possible reformation in the light of the scientific results to which it helps lead.

It must be noted that the 'metascience' approach remains influential, even where its methods are no longer those of deductive logic. For whatever the methodology, many philosophers still seek universal definitions of such terms as 'theory', 'explanation', 'scientific problem', 'scientific reason', and 'confirmation'. It has become popular to speak, in such connections, of 'criteria' or 'rules' – for example, of criteria of scientific explanation or problemhood or of being a theory or a scientific reason, or of rules of reasoning or of method – rather than of 'definitions' or 'explications'. But the criteria or rules are almost always conceived as being universal, and as neither discoverable nor alterable in the light of any results attained on the level of application of the criteria or rules to science itself. This sort of approach, which may be considered a third example of 'levels' approaches, is subject to a multitude of objections, many of them similar to ones raised against the approaches I have already discussed; for instance, what can be the justification of such criteria or rules themselves? Here, however, I will focus only on one objection, the inapplicability of such universal criteria or rules to all of science. Many such alleged criteria or rules have been criticized on the ground that they can be shown not to hold in specific cases from the history of science, so that they cannot be universally true of science.

Objections of these sorts, based on case studies, are in many cases well-taken. They have established that many prominent theories of the universal character of science are simply not universal. But there are grave dangers in the use and interpretation of such case studies, dangers which have not been guarded against sufficiently. Too many writers have concluded from their case studies that, if certain methods, goals, sets of rules or criteria, or whatever, do not hold for that particular case, they cannot be said to hold for any science at all. (Though such criticisms are directed against essentialist conceptions of science, they are themselves equally essentialist.) Similarly, some writers have tried to generalize, from characteristics they find in particular cases, to science as a whole. Neither of these extremes is defensible as an apprach to the understanding of science. If science changes not only in its accepted beliefs about the way things are, but also in its methods, goals, criteria, rules, and so forth, then what is found to be true (or false) of the science of a particular period may not be true (or false) of the science of another period. And of course what we need to understand is the way science *develops,* altering its belief-structure and its methodology, rules, criteria, goals, and so forth. The great contribution of recent studies of the history of science has been to help reveal that such changes do take place. That role of the history of science must continue, but the abuses of case studies, in serving as bases for generalizations about what science is and is not, must be avoided.[12]

One particular variant of the Levels Approach deserves special mention because of the appeal it has had for many philosophers. According to this type of view, the problems of philosophy of science are only special instances of more general philosophical problems concerning the nature of knowledge (epistemology) or concerning the nature of language; once those general problems are solved, the problems of philosophical interpretation of science will follow almost trivially. In epistemological versions of this approach, problems of the nature and role of observation in science are subsumed under problems about the nature of perception (sensation, sense-data, etc.)[13], and problems about the goals of science are seen as presupposing some general analysis of the meaning or criteria for use of the word 'know' or 'knowledge'. Philosophy of language versions claim that a clarification of such concepts as 'meaning' and 'reference' will settle issues about, for example, 'scientific realism' and the continuity of scientific reasoning from one fundamental theory to another.[14] In both approaches, a particular area of general philosophy, divorced from consideration of any specific scientific beliefs, constitutes the 'level' above and independent of science, and in terms of which the general features of science are to be understood. The long history of dismal failures to provide successful 'philosophical analyses' of perception, knowledge, meaning, reference, and related concepts should by now have made us at least suspicious that there is something *fundamentally* mistaken about such approaches – a point to which I will return at the end of this paper. Even apart from that point, modern studies of the historical development of science have shown how deeply and pervasively scientific change affects even our ideas of observation and knowledge: it becomes increasingly artificial to suppose that there *must* remain some common, essential features which it is the business of philosophy to exhibit. The impact of modern science in laying down the *Principle of Rejection of Anticipations of Nature* provides powerful additional support for abandoning such philosophical quests as quixotic.

For example, I have already mentioned how observation, in its role as evidence, can no longer be equated with perception in sophisticated scientific contexts, and the understanding of the role and interpretation of observation cannot be reduced to an understanding of perception – indeed, the two concepts are divorced almost entirely in such sophisticated contexts.[15] *Not only has traditional epistemology failed to provide the 'analyses' it promises; it turns out to have been misguided in principle in its methodological approach. For an understanding of the nature of knowledge – of the knowledge-seeking and knowledge-acquiring enterprise – can only be obtained through a study of the knowledge we have actually attained, of how we have attained it, and of how the goal of knowledge has itself been constructed and altered in that process.* I will discuss that process more fully in Part III below, and the faults of traditional epistemology more fully in Part IV.

Again, consider the concept of 'meaning'. That concept has long been notorious as one for which philosophers have consistently failed to provide a clear and adequate theory. Most often, such theories have approached meaning in terms of sets of necessary (and, for some theories, also sufficient) criteria or conditions for applying a particular term; the meaning of that term is declared to be given by that set of criteria or conditions, which *must* be satisfied if the term is to be applicable.[16] But the enterprise of seeking such meanings is misguided in principle, even apart from the failures of particular philosophical theories of meaning. What is important about scientific concepts is not any *common* features found in all its uses (much less the question of whether those alleged common features are necessary and unalterable 'analytic', i.e., essential ones). Rather, what is important are the ways those 'concepts' are *forged* in the light of beliefs which have been found to satisfy increasingly extensive and rigorous criteria of success and freedom from doubt, and the ways they are *changed* in the light of changes in such beliefs and associated changes in criteria. (It must again be emphasized that those criteria of success and freedom from doubt have themselves been forged in the same way, without vicious circularity.[17]) And the history of science brings out sharply, in a multitude of instances, how this has actually occurred. The most striking case is perhaps that of the chemical revolution, where new ideas and associated criteria of explanation and scientific goals ('compositional' as opposed to 'perfectionist' approaches to the understanding of material substances) led to a wholesale reconceiving, reclassifying, and renaming of material substances.[18] But more generally, philosophical theories of meaning, such as those which seek necessary and/or sufficient conditions of application of terms, cannot do justice to the development of such concepts as 'force', 'energy', 'field', 'electron', 'particle', or, indeed, any scientific concepts at all, including allegedly 'metascientific' ones. For in scientific cases, the emphasis is on developing concepts which are adequate, in the light of what we have learned, for thinking and talking about nature and for understanding it.

In order to deal with that emphasis, an entirely different approach is needed, one in terms of *reasons for change in science,* and in particular of reasons for change in 'criteria' (here, reasons) for applying a term. A clear notion of '(a) reason' in science, and of how and why that notion itself has evolved, is available, in terms of the notions of success and freedom from (specific and compelling) doubt.[19] In terms of that notion, a scientific 'concept' like that of 'force' or 'electron', or even 'explanation', 'reason', 'success', or 'doubt', can be characterized as a family of successive sets of criteria, each set in the series being related to its predecessors by the reasons leading up to its constituent criteria through changes in the constituent criteria of the predecessor sets, and to its successors by the reasons for changing its constituent criteria to those of its successors. (This characterization can easily be extended to relations of concepts which are contemporary with one another rather than of the 'ancestor-and-descent' variety.) We thus have a new concept of 'concept', one based on reasons rather than necessary and sufficient conditions.[20]

This characterization of what it is to be a 'concept' in terms of reasons is one which is, or at least ought to be, intuitively plausible for science. Science attempts to develop and consider all its ideas, 'concepts' as well as 'propositions', on the basis of reasons. This focus on conceptual change thus does far more justice to the attempt to understand the knowledge-seeking and knowledge-acquiring enterprise than do traditional essentialist theories of meaning, with their emphasis on conceptual constancy. The final burial of the latter sorts of view comes with the realization that, even if there are (as indeed there are in most cases) some features common to all the successive sets of criteria constituting a particular 'concept', the Principles of Rejection of Anticipations of Nature and of Internalization would lead to the rejection of the idea that those common features are essential, necessary. For they too ought to be open to the in-principle possibility of being altered in the light of new scientific reasons. Points analogous to those made in these last few paragraphs can be raised against most contemporary linguistic approaches in terms of 'reference'.[21]

## III

In the light of the preceding discussion of the lessons to be drawn from the failures of philosophical approaches to certain problems, and from the history of science, especially twentieth-century science, there are three main types of tasks which philosophy of science should be undertaking, functions which it should be attempting to fulfill.

### 1. The critical function of the philosophy of science

First of all, philosophy of science must continue its traditional task of exposing confused and mistaken interpretations of science. Such confusions and misconceptions often stand in the way of the development of a clear and adequate picture of science and scientific change, and therefore their dissolution is frequently a necessary step toward arriving at such a picture.

For example, one of the most important examples of error and confusion in the philosophy of science in the past three decades has to do with the problem of whether there is a 'given' in observation with which science deals. Following the full realization of the pervasive role of presupposition in science, even in the determination of what counts as observational and the interpretation of observation, many philosophers abandoned the notion that there is a given in experience. Standard philosophical criticisms of sense-datum and related theories of perception also contributed to the widespread acceptance of this viewpoint. This abandonment of the given in turn contributed to the relativism that became prevalent after the early 1960's; for if there is no given in experience, or if what we consider observational is shaped or determined by theory, then our views of the world are simply manifestations of our own (arbitrary) set of presuppositions, unconstrained by anything external to those presuppositions. But this sort of view neglected one crucially important aspect of observation in science, *an aspect which is tantamount to the existence of a given.*[22] Granted, it is not the 'given' as imagined in classical empiricist and positivist myth. It is not, that is, a given in the sense that it is found as a result of pure perception – perception purified, that is, of any prior belief. Much less is it a given in the sense that, once recognized, it can tell us automatically the character of the world by which it is given. Nor is it even necessarily the product of a single effort, a look, but is often the product of many efforts, complete with estimates of error. Rather, it is a given in the sense that, (1) having been marked out as significant by our best available background ideas, (2) having been appropriately described in terms of those background ideas, and (3) having been made accessible by application of background ideas (what I called in 'The Concept of Observation in Science and Philosophy'[23] a 'theory of the receptor'), the specific character or quantitative value we find it to have is independent of – not determined by – those background ideas. It is in that respect 'theory-independent'. This independent character or value, thus interpreted, must be taken account of by the theory which predicts a character or value of the thus interpreted factor, so that that character or value becomes the basis of a test of the theory. (It can in certain circumstances, often specifiable, also be a test of other background information entering into (1)–(3).[24])

Toward the end of Part II above, I proposed that science strives to employ as background information only ideas which have been found successful and free of specific and compelling doubt, and that such ideas count as reasons in science.[25] As an illustration of this, in (1)–(3) of the preceding paragraph, only such background information is, ideally, admitted. When we combine that analysis of scientific reasons with the above notion of the given, we get a picture of how science can develop rationally in the light of observational results. Whether we decide to distinguish 'reason' from 'observation' makes little substantive difference. If we do not choose to do so, then the given, appropriately selected, described (and, more generally, interpreted[26]), and accessed in terms of background information, simply counts as another sort of reason leading to scientific belief and change. If, on the other hand, we distinguish 'reason' from 'the given', then 'reasons' consist of the background considerations which have satisfied conditions of success and freedom from doubt, and in the light of which we select, describe, and gain access to the 'given' in observation. That 'given' can then lead to alterations in the background considerations in terms of which it has been selected, described, and accessed. (This second option with regard to the distinction connects it explicitly with traditional philosophical debates about the relative roles of 'reason' and 'experience' – here, observation – in obtaining and assessing beliefs and knowledge.) In either case, as we shall see more fully below, we have a clear and adequate way of understanding how scientific beliefs can be objectively arrived at, modified, and abandoned. And the path to such understanding has been prepared by the removal of a long-standing philosophical oversight.[27]

## 2. The overview function of the philosophy of science

Philosophy of science must provide an overview of the rationale of scientific change, of how science seeks and alters its beliefs on the basis of reasons, of why it considers some of those beliefs to be knowledge, and of how it has come to those conceptions of rationality and knowledge. In other words, this task consists in providing *a clear delineation of a general picture of science and scientific change which is compatible with the Principle of Rejection of Anticipations of Nature and the Principle of Internalization, and which avoids the confusions and errors of rival interpretations of science*. Such a general picture would provide a framework in terms of which the detailed investigations of the rationale of science, Function 3, can be carried out. That overview, which I have developed more fully in a number of other writings,[28] can be summarized as follows.

At earlier stages in the development of science or an area thereof, a distinction between what is and is not relevant to a certain conceived problem or subject-matter is rarely clear, or is understood in terms very different from those in which it may later be understood. The subject-matter – if there is a distinct subject-matter at all – may be distinguished in terms of certain obvious sensory properties, or in terms of some locality in which examples of the subject-matter are found, or in terms of some relatively simplistic analogy, or in a multitude of other ways. In some cases there may not even be a distinctive subject-matter, as with the Milesians' interest in trying to explain everything in general (or nothing in particular). Conceptions of what a (or the) problem is, and of what an answer to such a problem should be like (what, for instance, an explanation should be like), and of the means of obtaining such answers, may be based on any of a wide variety of considerations, if any at all; usually there is a great deal of vagueness in such conceptions, and sometimes there is no evident conception at all (e.g., of the means by which answers are to be found). Ideas of goals of such inquiry may also be vague, as may be ideas of what it would be to fulfill those goals (success) and to fail to do so (what counts as a doubt). To the modern mind, even calling some of the activities which are ancestors of science 'inquiries' may seem an outrageous stretching of that word's meaning (as with speaking of alchemical attempts to bring matter to 'perfection' as 'inquiries'[29]). Finally, at such early stages, there is often wide disagreement (the precise points of which may not be clearly understood) about the nature, description (interpretation), and relative importance of subject-matters, problems, goals, etc.; and there may be no clearly conceived and agreed-upon way in which such disagreements might be resolved.

But in the light of such goals, methods, problems, conceptions of success and failure, and so forth – however implicit and vague they might be – we have, *as a matter of fact,* arrived at further beliefs. Some of those new beliefs satisfied goals (standards of success and freedom from doubt) in terms of which they were sought, and some made those goals and so forth clearer, leading, for instance, to the selection of one interpretation of the previously vague and ambiguous goals. But in other cases, the beliefs arrived at led not only to alteration of prior substantive beliefs, but also to alterations in the goals ('So *that's* what we should be trying to do!') or in the methods ('So *that's* how to do it!') or conceptions of problems ('So *that's* the right way to see the problem!'), etc., which had been held previously.[30] All this is of course possible on the view presented here: for there is, after all, a 'theory-independent given', i.e., a 'given' which is independent even of the prior methods, goals, etc., which have led to the uncovering of a particular 'given' which can then lead to alterations of prior ideas.

Those reformulated goals and methods (etc.) in turn lead to new beliefs, which in turn may produce further alterations of the goals and methods as well as of prior substantive beliefs. A dynamic interaction thus develops between component elements of particular inquiry-situations. Some of the most important of those elements are, as we have seen: substantive beliefs, including beliefs about what to investigate or explain; the background beliefs which shape those substantive beliefs; ideas of the problems requiring such investigation, the methods by which to proceed, and the goals to be achieved. All of these elements may, as I have said, be quite ill-formulated, if formulated explicitly at all, in any given situation. Such developmental interaction has, over the history of science, gradually and often fitfully led to the following results of central importance for an overview of the scientific enterprise:

1. There has been an accumulation of beliefs, of various levels of generality, which have gained acceptance as having shown themselves *successful and free of specific and compelling doubt.*[31]
2. Associated with that accumulation of beliefs, a distinction has arisen, of generally increasing scope over the history of science, between considerations that are *relevant to a particular subject-matter or inquiry (considerations which are 'inter-*

nal' to science) and ones that are irrelevant thereto ('external' considerations).

3. The set of successful and doubt-free beliefs relevant to all scientific subject-matters (the total body of internal considerations) has shown itself to be *increasingly sufficient, with the increasing accumulation of such beliefs, to resolve problems regarding scientific subject-matters, without appeal to 'external' considerations.*

4. A body of conceptions of scientific methods, goals, and criteria of success and failure (doubt) has arisen; and, increasingly as internal considerations have become more and more nearly sufficient for the further development of science, these conceptions *have developed in response to such considerations* – even though the successive generations of ancestors of those conceptions of methods, goals, and so forth helped lead to those beliefs.

5. The subject-matters dealt with by science have themselves been reorganized and reconceived (redescribed, reinterpreted) in conjunction with the development of other elements of the scientific enterprise. The same is true of the scientific problems regarding those subject-matters and the relations between initially distinct subject-matters. (This fifth point will be discussed from a more revealing perspective in connection with Function 3, below.)

It is important to realize that all five of these results are contingent: there is no way in which their development could have been shown in advance, by *a priori* or transcendental arguments, to be a (or the) necessary outcome of inquiry. This is, of course, completely in accord with the Principle of the Rejection of Anticipations of Nature. The Overview itself describes a historical outcome which is contingent, and which could not have been anticipated by any *a priori* or transcendental argument purporting to show that it would or must be the outcome of inquiry; it had to be learned. (In particular, the Overview does not constitute an *a priori* 'form' of scientific inquiry and change, but is itself a product of that thought and is subject to change in the light of further scientific inquiry – recall the Cartesian room.)

Furthermore, in the case of all five results, debate is still possible, at least in principle. Doubts can exist regarding whether certain beliefs have actually shown themselves successful and free of specific and compelling doubt; whether certain considerations are or are not relevant to a particular subject-matter or inquiry (i.e. whether they are 'internal' to science); whether the body of successful and doubt-free beliefs is or is not sufficient to guide science in a given problem-situation; whether the body of methods, goals, and criteria which have emerged are fully acceptable; and whether subject-matters have been rightly conceived. Even if reasons for doubt do not exist or are themselves debatable at a given epoch in science, such reasons might in principle arise in the future. And such doubts of *specific* beliefs, if sufficiently pervasive, can in principle translate into doubts of the validity of the *general* results stated above. On the other hand, we must not forget that, in the case of a great many beliefs held at a given epoch in the history of science, there may be no specific reasons for doubt (or for holding those beliefs not to be successful); and furthermore, whether such doubts ever do arise is itself contingent; they may in fact never arise. Such beliefs, successful and free from specific doubt, serve as the grounds upon which we undertake new inquiries and arrive at new beliefs. They are indeed the only rational grounds we have for doing so. They alone (despite their vulnerability to the possible future rise of doubt) count as, and deserve to count as, knowledge (until and unless shown not to be).

Finally, it is only in the most general sense that these five results can be held to have been even *desirable* goals of inquiry, a desirability that could have been discovered in advance of actual inquiry, on the basis of common sense, everyday experience, or pure reason. That general sense has to do with the fundamental sources of human inquiry: of fear of the unknown and unpredictable, of desire and need to control the forces that affect us, or simply of curiosity. In a general sense, the desire is for the assurance of a body of successful and doubt-free beliefs which can be our safe guides through life; or it is simply for 'understanding'. These aims have to a large extent, and as a matter of contingent fact, been realized. But prior to inquiry, they could only have been formulated in ways so general as to provide little guidance, even as to what is desirable. And in any case, very different outcomes of inquiry might well have proved more desirable.

The description of the Overview results partly from a critique of misconceptions of science (Function 1), but primarily from a full realization of the two lessons of modern science, the Principle of the Rejection of Anticipations of Nature and the Principle of Internalization. It is thus itself a product of the enterprise of knowledge-seeking and knowledge-acquiring, not the result of an *a priori* inquiry.

But a full detailing of an Overview of the scientific enterprise must go well beyond such mere description. It must also, first of all, dispel doubts about whether it is itself adequate to account for the achievements of science. This is particularly necessary because traditional philosophy of science, like philosophy in general, has supposed that there are only two mutually exclusive classes of answers to the problem of knowledge: either there are certain, indubitable foundations or delimiting concepts of science (knowledge), or else we must succumb to relativism, subjectivism, or skepticism. Since the view of science which is compatible with the two principles cannot admit any certain and indubitable foundations or delimiting concepts, and yet must incorporate the possibility of knowledge, therefore a full development of an Overview of science which is both compatible with the principles and yet not committed to relativism, subjectivism, or skepticism must deal with questions of the following sorts:

- Why can the 'internal considerations' which become increasingly sufficient as bases for building further scientific ideas be considered to be *reasons,* and as the bases for scientific *objectivity?*[32]
- If the distinction between, on the one hand, considerations which are internal to science and count as reasons for or against scientific beliefs and as rational bases for further scientific development, and, on the other hand, considerations which

are 'external' to science, is a distinction which must itself develop, is there any basis for saying that that development is itself rational? Or would it be viciously circular to claim that the development of a body of considerations which will count as reasons is itself rational?[33]

– Can science have gotten off on the wrong foot at some initial point of its career, or, in building on what it supposes it has learned (beliefs which are in any case open to the possibility of doubt), might it have taken a wrong path, gone astray, somewhere along the line?

– Why, or in what respects – if at all – is science a better account of things, or of our world of experience, than any other – say, creationism, or astrology, or the Ashanti world-view, or the products of the Talmudic or Catholic religious traditions?[34]

– What is it that we have when, through the scientific enterprise, we obtain beliefs which are successful and free of specific and compelling doubt? (In the terms of traditional philosophy, what are the relations between reason-based beliefs and knowledge, between 'reasons' and 'truth'?)[35]

– What are the roots of the scientific enterprise – its sources in everyday experience and immediate human needs and assumptions, sources from which the enterprise departed, refining those needs and assumptions to produce new conceptions of methods, goals, *etc.*, as well as substantive beliefs?[36]

– How do approaches in science – methods, goals, etc. – achieve a normative status – a status such that certain methods, for example, *ought to* be employed, or that certain goals *ought to* be sought in science?[37]

Such questions, while their answers give deeper insight into the points included in the Overview, are dealt with in part by the kinds of studies described under Function 3, to which they form a transition. They can therefore be discussed in a more enlightening way in the context of that function, to which we may now turn.

*3. The detailing function of the philosophy of science*

The provision of an Overview of the scientific enterprise is without doubt an extremely important function of the philosophy of science. Its clear and incisive formulation is imperative as a means of providing an understanding of the overall character of science and avoiding confusions about that character. And it serves as a 'framework' for the construction of more detailed studies of the character of scientific change and its products. Nevertheless, it is such detailed studies that provide the deepest and most important insight into the scientific enterprise and its significance. A few examples of the kinds of detailed studies that are needed will enable us to bring this out.

– How have our present ideas of space and time (space-time) evolved from earlier ideas?

– How have ideas of matter and force and their relations developed from Kepler[38]

to supersymmetric generalizations of Grand Unified Theories in which fermions – the descendants of the idea of 'ordinary matter' – and bosons – the quantum fieldtheoretic carriers of force – are interconvertible?
- How did the view arise that explanation *must* be deterministic and precise (Plato would not have agreed), and why, and in what respects, did that view change, particularly in quantum theories? How has the idea of (what counts as) precision itself evolved?
- How, and in what sense, have we come to believe that symmetry principles, and the breaking of such symmetries, is central to fundamental physical explanation (i.e., of the origin and evolution of the universe and the entities in it)?
- How did the view arise that material substances are to be understood in terms of (i) their constituent parts, (ii) the arrangement of those parts, and (iii) the agencies (the forces) holding those parts together, and how has that view been modified since its early triumph? (Development of the 'compositionalist' approach to explanation of material substances.)[39]
- How did the piecemeal approach of scientific investigation develop – that is, the view that science should investigate particular areas of experience or nature (what I have elsewhere called 'domains'[40]) in isolation from other areas, and how has that approach been able to lead to unification of distinct domains which belies its own starting-point?[41]
- How has the view arisen that mathematics is applicable in the study of nature (Aristotle would not have agreed)? How has the body of applicable mathematics changed, and why?
- How has the concept of observation and its role as evidence in science evolved?[42]
Some of these questions are implicit in the discussion of scientific results leading to the Principle of Rejection of Anticipation of Nature, near the beginning of this paper, and that discussion will suggest other questions not mentioned here, and of course there are many, many more. Some, having to do with the internal structure of particular scientific theories, like those having to do with space-time and quantum theories, have for some time been concerns of philosophers of science, though their issues are transformed and become more tractable when expressed in terms of the Overview presented here.

The crucial point about all such questions is that they ask how certain centrally-important ideas of modern science – its most important substantive beliefs, methods, criteria, goals, and so forth – have developed. But they ask about such development in a particular way, a way which accords fully with the Overview resulting from the Principle of Rejection of Anticipation of Nature and the Principle of Internalization. They are concerned with how the ideas or approaches under study emerged from a background which was often vague and ambiguous, or which rested on implicit or explicit assumptions which would later be rejected; with how considerations relevant to the idea or approach under study came to be internalized into science, and with how earlier vaguenesses or ambiguities were removed (to the extent that they were), and with how assumptions behind the background came to be modified or rejected on the basis of such internal considerations; and with how those internalized considerations themselves interacted and altered, leading to further revisions and ultimately to the idea or approach under study. That is, *the concern of such questions is with the ancestry and descent of major ideas in science, relations which ultimately became unambiguously rational ones,* in the sense that they are based on relevance-relations which have shown themselves to be successful and free of specific and compelling doubt. That is, the questions are concerned with how those ideas and approaches are *rational descendants* of earlier ones.

This approach is profoundly different from that taken by the majority of historians of science today, who have come to be interested primarily in the history of scientific institutions, or in the cultural context of scientific developments. They no longer see the internal development of science as primary or even as clearly separable from the influence of allegedly 'external' factors contributing to that development. Indeed, their contentions are often tantamount to the denial of any important or useful distinction between external and internal factors in scientific change, a denial that has become popular partly because of the failure of previous philosophies of science to delineate a clear 'line of demarcation' between the two. That situation is altered fundamentally by the advent of the present view. For although this view abandons the misguided quest for a logical or essentialist line of demarcation, it explains how the distinction arises and plays a crucial role in the development of science: *it shows how the distinction can be methodologically usable by philosophers, historians, and sociologists of science.* (It does this irrespective of whether the evolved distinction is seen as having to do with 'rational' as opposed to 'irrational' or 'non-rational' factors in the development of science.) The distinction is a *de facto* one which can be denied only by an unrealistic view of how science proceeds. Furthermore, the present view implies that, while external considerations are important for a full understanding of science, nevertheless the independent functioning of internal ones in at least some cases (cases primarily of sophisticated science) can be ignored or denied only at the price of distorting the nature and accomplishments of modern science. Indeed, it implies that *any adequate portrayal of external factors and their roles presupposes the recognition and treatment of internal ones.* Let me elaborate on these points.

In interpreting the present view, we must guard against two possible misunder-standings. First, the concern with rational descendance does not imply that rational (internal) considerations are always in fact sufficient to guide scientific thought and work, even in its most sophisticated stages and areas. But when they are not, we should be able to recognize that fact: the very claim that such factors are insufficient presupposes that we can distinguish the rational (internal) ones. In other words, recognition of irrational or non-rational (or, more neutrally, external) factors presupposes that there are rational (internal) factors and that we can recognize them. Hence that sort of history of science that is concerned with external factors – and also sociology of science – depends on the kind of analysis of 'internal' factors discussed here.[43] Secondly, the concern with rational factors and their descendance does not deny the importance of non-rational or irrational factors for a full understanding of all aspects of science. Such factors can influence the development of science in many ways. They can provide a 'climate' in which certain sorts of ideas or approaches can flourish or wither; they can abet or hinder, encourage or block, the internal indications of science itself. *But that does not mean that there are not such internal indications:* those are what the 'external' factors abet or hinder, encourage or block, and sometimes the internal indications of science alter the external conditions – change society, for example.[44] There are indications given by the successful and doubt-free scientific background itself – sometimes clear, sometimes obscure; sometimes ambiguous, sometimes definite; sometimes debatable, sometimes conclusive. They have as a matter of contingent fact become more and more nearly sufficient for the guidance of science in its research, so that a study of internal development in such cases is possible. (More precisely, an internal study is possible which departs from examination of earlier stages in which the 'external-internal' distinction is either not clear, or is debatable and, in general, debated at that stage, or is formulated in terms quite different from those found at later stages.) It is with such internal aspects of science, and the development of such aspects, that the philosophy of science is, or rather needs and ought to be, concerned.

But such studies are important not only for delineating the scope of sociology of science and other disciplines concerned with 'external' as well as 'internal' factors in science. *They are important primarily because it is through such studies that we gain real understanding of the knowledge-seeking and knowledge-acquiring enterprise.* They provide detailed insight into questions of the general form, for what reasons have we come to believe the strange (and usually unanticipated) things we have come to believe? And why do those considerations, those 'reasons', count as *reasons?* And such studies are not restricted to the rational descendance of substantive beliefs: as my list of examples shows, they are also concerned with explanatory goals (as in the case of the question about the development of the compositionalist approach to explanation of material substances) and methods (e.g., the question about the rise and subsequent fortunes of the piecemeal approach).

Indeed, studies of these sorts having to do with substantive issues cannot be sharply separated from ones having to do with the rational descendance of what used to be called 'metascientific concepts'. That inseparability is a direct consequence of the internalization of those 'metascientific concepts' into the scientific process: of the historical (and contingent) fact, embodied in the Overview, that they have become so related to substantive beliefs that they too are modifiable or rejectable in the light of changes in such beliefs, and that alterations in them affect the body of accepted substantive beliefs in turn. Such interdependence is clearly exhibited in some of the above sample questions, for instance those involving changing criteria of explanation.

That interdependence brings out the extent to which the threefold division of functions of the philosophy of science is a rather artificial one. We have already seen how intimately Function 1, the Critical Function, is related to Function 2, the Overview Function: a clear and convincing statement of the Overview of the scientific enterprise must often involve criticism of incorrect but influential and sometimes plausible pictures. Demonstration of the existence of a 'given' also involves Function 3 tasks, of detailing in specific cases how such a 'given' is selected and interpreted in terms of well-grounded background beliefs, and how background beliefs show us how to gain access to it.[45] (Some Function 1 tasks, however, consist simply in removal of conceptual or other types of confusion.) Now we see how Function 3, the Detailing Function, also is intertwined with Function 2, the Overview Function. For in getting a detailed view of the rational development of such concepts as explanation and observation – always in close conjunction with a detailing of alterations in associated substantive beliefs – we get insight into the kinds of considerations that serve as reasons in particular but often complex problem-situations. And we get insight into why they serve as reasons. For example, 'The Concept of Observation in Science and Philosophy',[46] by bringing out in detail the various exact roles played by successful and doubt-free background ideas, shows why certain features of a particular case count as observational evidence for or against certain hypotheses, those hypotheses themselves being constructed in the light of such background information. (That essay also performs the Critical Function in showing how certain standard views of the concept and role of observation are mistaken.)

A detailing of the concept of 'reason' itself, of the same sort as other Function 3-type studies, can be provided. Such an analysis is offered in 'Objectivity, Rationality, and Scientific Change',[47] where central aspects of what it is to be a reason in science are traced to the rise and triumph of the piecemeal approach to inquiry (one of the examples given above of a Function 3 question). The piecemeal approach, it will be remembered, approaches particular subject-matters (domains) in isolation from others; its possibility is contingent, since things might have been such that, say, the forces of nature (or 'internal relations between all things') were so strong that no things or events could be studied without taking into account the effects of other things or events. However, the piecemeal approach did prove possible, and through it science has been able to achieve understanding of a great many domains, and has even been able to transcend the piecemeal approach itself, arriving at unification of the accounts of many disparate domains. In connection with the adoption of the piecemeal approach, the ideas of 'success' and 'freedom from doubt' were (as I go on to argue in that paper) provided new and far more precise and developable criteria than before, although those new criteria can be seen to have been rational descendants of prior ones.[48] This furnishing of new criteria came about through the association of those two ideas with a structure of categories of scientific problems brought to the fore by the piecemeal approach. Consequently, what counted as legitimately employable background beliefs for guiding scientific inquiry was restricted to ones which had proved successful and doubt-free *in the sense given by the piecemeal approach.* That sense was later adapted, with appropriate modifications, to the increasingly unified science that evolved as a result of the piecemeal approach. Furthermore, those new criteria of 'success' and 'freedom from doubt' are related to (are rational refinements of) certain everyday ideas of what counts as a 'reason', ideas which themselves have their roots in everyday circumstances and needs of human beings.

Thus the rational descendance of the concept of 'reason' itself is given in the context of a Function 3-type study. That is, through such a study, we obtain a partial answer to the first in the list of Function 2 questions having to do with potential doubts about the Overview, the question of why internal considerations count as 'reasons' in the scientific enterprise. (The paper summarized in the preceding paragraph also gives an account of what is involved in the 'objectivity' of science.) We also get a partial answer to the second question, the question of how the concept of '(a) reason' – of what counts as a reason – in science can itself be said, without vicious circularity, to evolve rationally. Briefly, that partial answer is as follows: at each stage of development, there were criteria or standards of rationality – 'rational' because descended from fundamental everyday human circumstances which give rise to the need for a concept of 'rationality' for standards (however vague and ambiguous in origin) which would distinguish 'rational' from 'non-rational' or 'irrational' thought, inquiry, and action. And the criteria developed were, at various succeeding stages, clarified and altered in the light of what was found to be believable and achievable. In the case of both of these Type-2 questions, a full treatment must also attend to the the Critical Function of dispelling misunderstandings arising from the philosophical tradition and elsewhere. (Note that the sixth Function 2 question, regarding the roots of the scientific enterprise, is also touched on in this account.)

That paper, 'Objectivity, Rationality, and Scientific Change', also shows the way to deal with the third Function 2-type question: of whether science, precisely because it relies so heavily on antecedently successful and doubt-free beliefs, might have gone off on a wrong tangent at some point in its history. The answer is that it indeed could have: all our beliefs are subject to the possibility of being wrong; there are no guarantees. But the *mere* possibility that we may have gone wrong somewhere is not itself a reason for doubt. It is not, that is, the sort of reason for doubt that plays a role in the actual knowledge-seeking enterprise – one which counts for or against a specific belief. It is the sort of spurious 'doubt' – a philosopher's doubt – that holds for *any* belief whatever, including its own negation; it is tantamount only to the recognition that specific doubts can always arise with regard to any specific belief. Therefore if we have gone wrong somewhere, we can only hope that specific reasons will be revealed which will expose our error, and that we will then be able to act accordingly to correct our mistake. Otherwise the mere fact that we might be wrong is no reason to doubt beliefs that have proved successful in accounting for their domains and free of *specific* doubt.

Other Function 2 questions require, in addition to such detailing, other or further modes of treatment. This is particularly the case with the question, 'Why or in what respects – if at all – is science a better account of things, or of our world of experience, than any other view?' Rival views must be dealt with on a case-by-case basis; nevertheless, most of the purported rivals fail to match science on one or more of the following general sorts of grounds. Some of them fail to fulfill their own promises, their own goals, their own standards of success: astrology promises predictions of future events on the basis of planetary 'aspects'; but it fails to deliver. (Auxiliary hypotheses to explain the failures do not help, since, among other things, they can explain particular failures only after the event, thus betraying after all their initial promise to predict.) Others provide no way of answering their own questions: in at least some of its forms, creationism gives no way to decide whether the universe was created six thousand years ago, or six million or fifteen billion years ago or six seconds ago; it merely asserts that what we observe is compatible with any creation date. (Of course one could accept a particular date as a revelation to be accepted on faith; but to do so is, *in terms of the faith view itself and not of some competing view,* an explicit rejection of the idea that there is any reason at all for the belief.) In neither of these sorts of judgments are we appealing, in our criticisms of rivals to science, to some supercriterion of judgment. For example, to say that a theory fails to fulfill its own goals, or that it fails to satisfy its own criteria of success, is not to appeal to a further criterion outside and above that theory. ('A theory should fulfill its own criteria of success' is not a transcendent criterion which might be denied by a particular theory under scrutiny; it is merely a general way of saying that, on its own internal grounds, the theory in question is not a good theory.) Some other theories are parasitic on the results of science, claiming to explain exactly what science explains, but by different assumptions: the theory that we are all brains in a vat, being programmed to think that things are a certain way when they are not, accounts for the very same statements that science makes (perhaps along with some others). The trouble with such theories is that they are equally compatible with the denial of all the statements they purport to explain, and so do not distinctively explain at all; at best, they are *mere* possibilities, mere conversions of philosophical 'doubts' into a positive form of expression. (The fault is badly stated when it is said to lie in the unverifiability or unfalsifiability of such views, as if it had something to do with testing them. Their real flaw lies in the betrayal of their own promise to explain. Other examples of such 'rivals' are little-green-men explanations and some versions of creationism, including the one just discussed.) Again, no supercriterion, transcending the theory or view under scrutiny, is being appealed to. In some cases, the very fact that we have *found it possible* to give precise and detailed accounts of experience (and 'things' in general) makes science preferable to views (like the Platonic) in which *less is asserted to be possible* in explanation. (We have found that more can be achieved than Platonism says can be achieved. Again no supercriterion is presupposed.)

In yet other cases, the question of whether a rival view or theory or approach is 'better' than science requires close attention to the various senses of 'better': better for gaining knowledge, better for helping people be 'happy', better for helping them be 'moral', and so forth. Making such distinctions requires no assumption of a point of view superior to either science or the theory in question; it can be accomplished by examining the goals of that theory, or the characteristics of its domain.[49] Only with such distinctions in hand can the question be taken up of whether the rival bests science for the purpose of making people 'happy', for instance – as well as of whether *that* question is answerable independently of the question of which is 'better' as a body of or means to knowledge.[50]

That the detailing function of the philosophy of science is conducted within the framework of the Overview does not imply that the Overview itself is not subject to modification in the light of the cases whose selection and analysis it guides. On the contrary, we have been taught the Overview by the lessons of the history of science, and of modern science especially, which has instructed us in the Principles of Rejection of Anticipation of Nature and Internalization. As the Overview has thus been learned from the concrete events of the development of science, so also it is, *like any scientific theory, whether of nature or of how to learn about nature,* subject to revision. We have learned to make such revisions in the light of scientific results, including our study of those results according to the procedures described in the present section.

## IV

The revisability of the Overview of scientific change portrayed in the preceding section is a further demonstration of one of the central tenets of the present view: that science has developed an intimate interaction between the beliefs at which it arrives and the methods by which those beliefs are attained. Even our view of science and its methods, and of the methods by which *that* view is arrived at, are interlocked with the rest of science. Philosophy of science does not exist and function on a level above and independent of the substantive content of scientific beliefs; it is integrally and inseparably linked to that content, and its methods and conclusions must rest on the results of the very science with which it is concerned.

This view of the philosophy of science and its relations to science has certain affinities with what is called a 'naturalistic' approach to the theory of knowledge, as advocated by Quine and others.[51] Quine sees that an understanding of our knowledge-seeking and knowledge-acquiring processes, and of the validation of the results of those processes, must rest on the results of science itself: 'the epistemologist may make free use of all scientific theory'[52]; 'Why not just see how this construction [of a picture of the world] really proceeds?'[53] But when, in the sentence following this rhetorical question, Quine spells out his recommendation, it turns out to involve not the free use of *all* scientific theory, but only of psychology: 'Why not settle for . . . a surrender of the epistemological burden to psychology?' 'Epistemology, or something like it, simply falls into place as a chapter of psychology and hence of natural science'.[54]

Quine's reasons for considering epistemology to be a branch of psychology seem to be as follows. Knowledge, he holds, consists in 'working a manageable structure into the flux of experience',[55] where 'experience' is equated with 'surface irritations and internal conditions'.[56] Since psychology is concerned with sensations and the 'internal conditions' which determine their interpretation, it follows that philosophical inquiry about knowledge is to be approached through psychology. In the theory of knowledge, 'The relation to be analyzed, then, is the relation between our sensory stimulations and our scientific theory formulations: the relation between the physicists's sentences on the one hand, treating of gravitation and electrons and the like, and on the other hand the triggering of his sensory receptors'.[57] And so the object of the epistemologist's study is 'a natural phenomenon, viz., a physical human subject'.[58]

Now, psychological considerations play no role *as reasons* for raising specific problems, employing specific methods, or accepting or rejecting specific beliefs in any actual science (except perhaps psychology itself) – at least in sophisticated science. This is abundantly clear in the solar neutrino experiment: the considerations that are brought to bear as reasons in the conception and performance of that experiment, and in the construction of hypotheses to account for its negative results, are specific propositions, for example, ones taken from the theories of stellar structure, stellar evolution, and nuclear physics. Psychological considerations occur nowhere in the reasoning (except insofar as they are guarded against as potential dangers to the accuracy and objectivity of the experiment). Indeed, they could not, at least at the present stage of the development of science: psychology simply does not have that sort of specific relevance to *most* scientific areas.

But it is also the case that psychological considerations play no role in helping us understand *why the considerations that actually are employed count as reasons*. Again referring to the solar neutrino experiment as an example, we find that what makes those considerations count as reasons ('background information') is that (1) they have been found to be *well-grounded (successful and doubt-free)*, and (2) they can be shown (or plausibly argued) to be *specifically relevant* to the problem being dealt with in the experiment.[59] And what is true of the solar neutrino experiment is true of sophisticated science in general: in such science, *the sorts of considerations that are relevant to the formulation of problems, the introduction of new ideas and the reformulation of old ones, and to the acceptance or rejection of ideas, are considerations that have been found to be well-grounded and relevant in specific ways to the specific subject-matter in question.*[60] It is such considerations, together with the data given in observation (in the sense of 'given' outlined earlier) that count as reasons in science.

Thus, to understand the concept of '(a) reason' as that concept is employed in modern science, we need to attend to the ways in which ideas (and also problems, methods, possibilities, goals, etc.) are accounted successful and doubt-free – as potential background information – and how such ideas (etc.) come to be accounted relevant to specific problem-areas. To this task psychology is irrelevant. What *is* relevant is the examination of the way scientific reasoning itself proceeds. We must take seriously (as he himself did not) Quine's own question, 'Why not just see how this construction [of a picture of the world] really proceeds?'[61]

Can psychology at least illuminate the rationale of the scientific process through what it has to say about human sensory receptors and their 'triggering'? No. We have seen that, as the scientific process becomes more and more sophisticated, more and more autonomous in its modes of argument, the relevance of sensory input to inquiry and its products becomes less and less fundamental to that process, more and more derivative in its role. For example, as we learn more and more about what sorts of information to seek as evidence, that information – its selection and interpretation, and the appropriate ways of gaining access to it – becomes more and more dependent on well-founded background ideas (as, again, in the solar neutrino case). Sensory input and its triggering becomes increasingly peripheral to the evidence-gathering process. Even the 'given' need not be a sense-perception, and though we must at some point be made aware of it, our awareness of it is not what is given, and does not constitute the interpretation of the given as evidence. Indeed, as is brought out precisely in the neutrino case, we have learned the necessity of taking care to keep psychological considerations, including reliance on human perception for evidence, out of the picture, as being inadequate as evidence, and as being a source of potential interference with the objective conduct and assessment of the experiment and its results. Human sensation is no longer epistemically primary in the knowledge-seeking enterprise, at least in sophisticated science: as far as human sensation is concerned, what have become epistemically important are the constraints on its use and interference. As far as evidence is concerned, other sorts of considerations, having to do with well-grounded, doubt-free, and relevant beliefs, have assumed the central role.

It is certainly very possible that someday we may find that certain aspects of human psychology are relevant, in specific, well-grounded, and well-understood ways, to the epistemology of the knowledge-seeking and knowledge-acquiring enterprise and its assessment. But as matters stand today, there is no such relevance. And considering the pervasive disagreements among psychologists about fundamental psychological theory and methodology (they do not even agree about the proper vocabulary for talking about their subject-matter), and the quite primitive state of many areas of psychological investigation, perhaps this is just as well. If it were relevant, the best advice we could follow would be, 'Wait; save your epistemological questions until (much) later'.

In summary, Quine's conception of the problems of epistemology is flawed in two fundamental ways. First, in speaking of 'the relation to be analyzed' as one between 'the physicists's sentences' and 'the triggering of his sensory receptors', it calls attention away from the real reasoning employed in constructing scientific problems, methods, and ideas, and in reaching, interpreting, and testing scientific conclusions. As we have seen, that reasoning has to do, in sophisticated science, with the employment of ideas that have been found to be successful and doubt-free, and whose specific relevance to the problem at hand has been shown or can be argued for (again on the basis of well-grounded background beliefs) as plausible. Of course, the 'relation' could be spelled out in terms of the roles played by such ideas. But Quine shows no inclination to do this, and in any case, if it were so spelled out, epistemology would no longer be the study of 'a physical human subject', but rather of something more appropriately characterized as the investigation of relations between propositions.[62] It is science, not the people who do science, that shapes further science – at least after a critical mass of internalized background information has accumulated.

Secondly, Quine's focus on 'the triggering of sensory receptors' as an ultimate term in the relation to be studied also fails to deal with the real issues about the scientific process and its results. What counts as observational, as evidential, in science – and, correspondingly, the role of sense-perception in epistemic contexts – is as subject to revision as any other scientific idea. And changes – profound changes – in these respects have taken place: the rational descendant of 'All our knowledge of nature is based on sense-perception' is 'All our knowledge of nature is based on interaction with nature', where 'interaction' is specified by a particular body of well-grounded scientific claims.

Both these flaws stem from a deeper one: Quine's lack of concern with actual scientific thought, an omission that leads him to serious misinterpretations of the scientific enterprise, *misinterpretations that ignore the major lessons and achievements of modern science*. Consider, for example, such statements as this: 'The quest for knowledge is properly an effort simply to broaden and deepen the knowledge which the man in the street already enjoys, in moderation, in relation to the commonplace things around him.'[63] Such statements fail to recognize the ways in which science has departed radically from the views of the 'man in the street'. They fail, that is, to grasp the import of the Principle of Rejection of Anticipations of Nature, which we have seen to be one of the major lessons of modern science. Or, again, consider the following: 'The scientist is indistinguishable from the common man in his sense of evidence, except that the scientist is more careful. This increased care is not a revision of evidential standards, but only the more patient and systematic collection and use of what anyone would deem to be evidence'.[64] Does not this view constitute a failure to recognize the extent to which science has managed to alter its very concepts of evidence, its very methods, its very aims, to build its reasoning-processes on what it has learned, rejecting, in the process, the evidential standards and reasoning-patterns of the common man who anticipates nature? Is it not a failure to grasp the lesson of the Principle of Internalization, a principle by which science builds on what it has learned, developing means of choosing rationally, in some cases at least, between alternative possibilities?

That epistemological problems about scientific knowledge must be approached in the light of the substantive scientific knowledge or knowledge-claims we have available has been a deep and important insight. But far from having seen how to carry out such a program, Quine has continued to appoach those problems in ways that fail to make use of such knowledge. Usually it is purely logical and linguistic considerations that provide the bases for his claims about science, claims that are wholly independent of the specific substantive content of the sorts of scientific beliefs to which the claims are relevant. His conception of the role of psychology is of the same sort: it stems from a conception of a completely general (alleged) feature of science: its concern with sensory input, a concern that is again completely independent of any specific scientific conclusions.

Such failures to look deeply at the reasoning involved in modern science, and to grasp the major lessons of that science, have disastrous consequences for the interpretation of the knowledge-seeking and knowledge-acquiring enterprise. Consider Quine's Duhemian thesis that 'Any statement can be held true come what may, if we make drastic enough adjustments elsewhere in the system'[65], that any part of the total system can be preserved in the face of any observation. This view is not the result of an investigation of actual scientific reasoning. It is in fact a purely logical point, that if more than one proposition is necessary to deduce an observational prediction, and if that observational prediction is not borne out, then any one of the premises may be at fault, and any one might be salvaged by putting the blame elsewhere. Quine is aware that scientists often have quite definite and specific grounds for accepting one modification rather than another: 'The scientist does indeed test a single sentence of his theory by observation conditionals, but only through having chosen to treat that sentence as vulnerable and the rest, for the time being, as firm'.[66] But the important point, for understanding the knowledge-seeking enterprise – the important epistemological issue – is whether there are reasons for making that choice: why, and under what circumstances, the scientist considers one hypothesis to be more doubtful than others. And it is clear that in many cases in sophisticated science, such decisions are made, and that the decisions are made on the basis of reasons: on the basis of considerations which have proved successful, doubt-free, and relevant to the problem at hand. Quine, however, never enters into a discussion of what those reasons and circumstances might be.

But if any statement can be maintained come what may, and if we are given no analysis of the sorts of reasoning by which we might choose between the logically open alternatives, the result can only be relativism. The pretensions of science to knowledge are unfounded; any view can be held if we are clever enough. Other Quinean doctrines – the indeterminacy of translation, the underdetermination of theories by experience – also contribute to this relativism, and though their examination is beyond the scope of this paper, the question must at least be raised as to whether they too would look very different in the context of a closer examination of actual scientific reasoning.[67]

Quine is far from alone, either in his failure to approach questions of the theory of knowledge and the philosophy of science through an examination of science, or in the relativism and skepticism consequent on that failure. He has at least seen the necessity of approaching problems about the knowledge-seeking enterprise through an examination of actual scientific results, even though he has not succeeded in doing this as it needs to be done. Most of his philosophical colleagues have not even gone that far. The view that the scientific enterprise can and indeed must be approached through some 'level' independent of any substantive scientific beliefs still prevails in philosophy – the idea that, in order to understand reasoning and knowledge (about nature), we must provide *absolutely general* characteristics or criteria which apply to *all possible* reasoning and knowledge, and are therefore independent of *any particular* cases of reasoning and knowledge. On that assumption, the evolution of science itself, with its developing modes of thinking about nature, is irrelevant to epistemological questions, and the latter must be approached through methods or disciplines which are independent of any particular views of how and what we know. The methods which have been thus employed by philosophers are familiar; among them are transcendental arguments and 'conceptual analysis' or analysis of 'meanings' (or of the general concept of reference); among the disciplines relied upon are logic and the philosophy of language. The nearest such approaches ordinarily get to the application of scientific results are a few generalities taken from linguistics and psychology; but even then little or no attention is paid to substantive results in other areas of science and the way they might affect and illuminate the knowledge-seeking enterprise. Even in Quine we see this influence perniciously at work: the opportunities suggested by his naturalistic epistemology are distorted into a view of psychology as applicable to inquiry about knowledge in the same old 'levels' sort of way. On the one hand, his heritage as a logician leads him to concentrate on those aspects of reasoning that have to do only with logic (as in his Duhemian thesis), ignoring the vast body of additional reasoning which scientists employ in obtaining results whose attainment formal logic alone cannot sanction. And on the other hand, as we have seen, the rationale behind the epistemological importance he imputes to psychology lies in the assumption of the *general* and fundamental relevance of sensation to science. In this light, Quine's conferral of the tasks of epistemology on psychology appears as the outcome of a residual commitment to traditional philosophical conceptions, in particular to the view that science is constructed on (or tries to order) sensation, and that analysis of what is involved in sensation is therefore tantamount to an analysis of what is involved in the seeking, acquiring, and assessing of knowledge.

That traditional philosophical view is, however, a dogma – no, an error – of classical empiricism, wholly inadequate for illuminating the scientific enterprise.

And that dogma or error is only a special case of a body of intertwined errors of broader scope and influence, common to most of traditional epistemology and philosophy of science: the errors of supposing, first, that the knowledge-seeking and knowledge-acquiring enterprise can only be understood through considering the most general aspects that belong to *all* knowledge-seeking and *all* its products (that there even *are* such general aspects is itself a Platonic dogma); second, that philosophy or some branch thereof (perhaps supplemented by a few general features of psychology or linguistics) has the task of finding those general aspects; and third, that that task, because of the very generality of its aims, must be accomplished independently of any specific methods by which actual knowledge-seeking proceeds, and of any specific results that have been attained by those methods. (That the general aspects can be content-independent is also a dogma.)

Such philosophical approaches have missed the deep lessons of modern science, the lessons of the Principle of Rejection of Anticipations of Nature and the Principle of Internalization. They have missed, that is, the way modern science has managed to incorporate allegedly independent 'levels' into its processes in order to test them. They have missed the ways in which humanity has been able to learn from its experience, the profound ways in which that learning-process has altered not only our views of nature, but also the ways in which we think about it and learn about it. They have missed the ways in which the scientific enterprise has altered what we count as reasons in our investigations, what we take to be evidence (observation), and what we take as the goals of our investigations; they have missed the ways in which we have been led to alter even our conceptions of what investigation consists in.

In the light of these omissions, we should not be surprised that skepticism and relativism are the common results of philosophical epistemology. Where its accounts of reasoning are relevant at all (as in the Duhemian problem), they are so impoverished by neglect of the full richness of scientific reasoning that what science claims to do (and has done) appears impossible, a delusion. Sterile in its methods, barren in its results, philosophical epistemology, including much of standard philosophy of science, has been unable to account for the enterprise of science and the results it has attained.

Richard Rorty has seen the consequences of the traditional epistemological approaches and has proclaimed them bankrupt.[68] In this he is correct. But Rorty has seen the issue only from within philosophy, as a result of purely philosophical inquiry, rather than as the consequence of ignoring the actualities of the scientific enterprise. He has therefore been led to equate his rejection of epistemology with a banishment of privileged knowledge-claims also, and therefore, in effect, to view the former as a judgment by philosophy on the claims of science to rationality and knowledge: there remain, for Rorty, only different points of view and 'conversations' between them.[69] But although epistemology in its traditional philosophical sense is indeed bankrupt, and with it much of what that subject proclaimed knowledge to consist in, that is not because science has failed in its attempts to find knowledge on the basis of reasoning, but rather because philosophical epistemology has failed to get at what reasoning and knowledge consist in. Far from being relevant to what science says, philosophical epistemology has withered because it has made science irrelevant to it.

Science in the twentieth century, building on prior results while altering them profoundly, has made tremendous strides in gaining an integrated insight, however incomplete and subject to future revision, into such grand topics as the basic forces of nature, the early stages of the evolution of the universe, the evolution of elementary particles and the chemical elements, the nature and development of life, and the future course of the universe. While all this has been going on, philosophers have been 'analyzing the concept of knowledge' and developing 'the logic of investigation' in complete oblivion to it, as if such facts were inadmissible evidence in the quest for universal criteria and rules governing the knowledge-seeking enterprise. And yet in the process of reaching its new conclusions and problems, science has been led to profound alterations of our preconceptions about the criteria and rules involved in seeking understanding and knowledge, and about what such understanding and knowledge must consist in. We are long past due in taking seriously those alterations and the reasoning by which they have been arrived at. A comprehension of how we rationally seek and attain knowledge-claims, and of the sense or senses in which they constitute knowledge, requires that we now turn our attention, in the ways delineated in this essay, to their study.

## Notes

1. Bacon (1939), p. 33.
2. Shapere (1986b), (1984).
3. Shapere (1984), especially Chs. 14, 18.
4. Shapere (1984), (Unpublished a), (1982).
5. Shapere (1986a), (1986b), (Forthcoming), (Unpublished b), Introduction to (1984).
6. Especially (1986a), (1986b), (Forthcoming), (1984), (Unpublished b).
7. Shapere (1986b), and below, Part III.
   I have discussed the necessity that the doubts be specific rather than universal (i.e., applicable to any belief whatever, including the denial of every particular belief) in many places; see, for example, the Introduction to Shapere (1984) and several of the papers in Part III of that book.
   In a review of Shapere (1984), Nickles (1985) views the distinction between specific and universal doubts as the 'core argument' of my position. As should be evident from the present paper, however (and indeed as should be clear from the Introduction to Shapere (1984)), it is very far from that. The distinction is important primarily because so many philosophers, from Descartes to the present day, have considered universal 'reasons' for doubt (like 'I may be dreaming', 'A demon might be deceiving me', or 'I may be a brain in a vat') as reasons for suspecting the results of science, or, more generally, for doubting that we have any knowledge at all. My point is that *such 'doubts' play no role in actual inquiry because, as a result of their very universality, they provide no selective guidance with regard to which ideas to accept or reject, or with regard to what directions should be taken in further inquiry.* To

have thought they did, and that therefore they could count as 'reasons', was and continues to be a philosopher's confusion; the only role that can be attributed to such 'doubts' is that they are misleadingly-formulated reminders that specific doubt *might* (could in principle) arise with regard to any specific belief we might hold, no matter how successful and free of specific doubt it may at present be. But the mere possibility that such doubt might arise is not itself a reason for doubting such a belief. (There is thus no ground for Nickles' worries as to 'precisely *how* science has learned [that such universal or philosophical doubts play no role in the scientific or knowledge-seeking enterprise].')

Dispelling such arguments is an important function of the philosophy of science (in the terms of Part III of the present paper, such dispelling falls under Function 1 of the philosophy of science, the Critical Function). But what is of vastly greater importance for the philosophy of science is the development of a *positive* view concerning the nature of scientific reasoning. It is this which forms the 'core' of the view which is outlined in the Introduction to Shapere (1984) and, in a more up-to-date form, in such works as Shapere (1986b), (1982), (Unpublished b), (Unpublished a), the present paper, and, in more detail, (Forthcoming).

Nickles also appeals to the currently fashionable argument from 'negative historical inductions' – to 'some disturbing lessons from the history of science itself, for example, that virtually every major theory is bound to fail at some point and that our conceptual framework in the year 3000 . . . is likely to be quite different from our framework in 1985 . . . do [such arguments] not furnish serious reasons for doubt?' Such alleged 'inductions', however, play very fast and loose with the notion of 'failure' of theories (and also with the notion of 'theory'), ignoring the fact that much (even of a 'theoretical' sort) has been retained from past science, and that even 'rejected' theories have in many cases been shown to remain applicable within limited domains. Indeed, an exactly opposite 'induction' – *equally or more plausible, but ultimately no less simplistic and misleading than that to which Nickles appeals (but no more so either)* – can easily be constructed. According to it, no theories have ever been rejected, past theories only having been shown to be of restricted application. (Aristotle's physics is not false; it has merely been shown by Newtonian physics to hold only within the restricted domain of a gravitational field with friction and air resistance operating.) Hence our current 'theories' can be expected to be retained in the same way and to a comparable extent.

But it is not necessary to construct such a counterinterpretation. For even on the most charitable interpretation of Nickles' inductive argument, the history of science does not show inductively that virtually every major theory is bound to fail sooner or later. On any measure of the quantity of science done, far more science has been done in the twentieth century than in all the time previously; and that quantity has increased rapidly during the twentieth century itself. Yet in all this effort, despite expectations to the contrary, some of the most important theories in twentieth-century science have continued to pass all tests. For example, quantum mechanics, originally found to hold at the atomic level, was expected by some (including Bohr) to fail for the nucleus; it did not. Later some scientists believed that it must fail as an account of the constituents of the nucleus, and, later still, that it, or its field-theoretic extensions, must fail for the constituents of those constituents. Despite such expectations and claims (sometimes resting on quite specific problems about the subject-matter), quantum mechanics and its extensions have not failed, but have proved able to encompass an ever-increasing body of information, from smallest levels of investigation to the universe itself and its origin. Though there have been important developments of the theories, those have had the character more of extensions than of replacements of quantum mechanics. Surely, in terms of the effort applied – the quantity of scientific work involved – any measure of such success (and, in a right-minded sense, accumulation) should carry greater inductive weight than the failures on which the opposite contention relies.

In any case, other than as an allegation of skepticism about current science, it is not clear how Nickles' induction is supposed to affect my arguments. For I am of course perfectly willing to admit that our present views *may* turn out to be incorrect, even though we have no specific reasons to suppose they are. But the fact that they *might* (in principle) turn out to be incorrect does not affect the point that, in the case of at least some of our current views, all the reason we *do* have favors them, and that it is therefore reasonable to rely on them.

Nickles claims that ' "Freedom from doubt" is surely too strong; and how does Shapere answer the usual objections to the pragmatists' epistemic use of success?' (p. 311) In Shapere (1986b), I have pointed out that beliefs which are not entirely free from doubt are often employed as reasons in science, and detailed the circumstances in which such employment is legitimate. But the plentiful availability of beliefs which *are* free from specific and compelling doubt, and the sufficiency of such beliefs as bases of scientific reasoning in a great many cases, must not be dismissed lightly. As to 'the usual objections to the pragmatists' epistemic use of success', both the objections and my answers to them are given in (1986b). It must be understood that that essay gives an analysis of the precise senses of 'success' and 'freedom from doubt' that are relevant in science, and of how those senses have been developed through the scientific enterprise. A brief sketch of the results of that analysis is given in Note 48, below.

In the present paper, I have not discussed explicitly the requirement that the successful and doubt-free considerations must also be *relevant* to the context in which they are used as reasons. The omission is not crucial, since satisfaction of the first two requirements – success and freedom from doubt – have to do with what can (potentially) be a reason in science; from the body of potential reasons, some can be reasons in the context of a particular problem by virtue of their established relevance (*cf.*, Shapere (1986b), (Forthcoming).

8. *Cf.*, Shapere (1986b).
9. Shapere (1986b).
10. Shapere (Unpublished b).
11. *Cf.*, Shapere (1984), especially Ch. 11, for further discussion of the topics of this section.
12. See Shapere (1984), Ch. 10, for further details and discussion of other aspects of methodology in the philosophy of science.
13. Shapere (1982).
14. Shapere (1984), Ch. 18.
15. Shapere (1982), (Forthcoming).
16. I discuss necessary-and/or-sufficient-condition analyses of 'meaning' only as an example, not to suggest that all traditional philosophical theories of meaning are of that sort. Those other sorts of philosophical (i.e., universal, and usually essentialist) theories of meaning have their own difficulties.
17. See Shapere (1986b) and Part III, below.
18. Shapere (1984), Chs. 15 and 18.
19. Shapere (1986b) and Part III, below.
20. Such families of reason-related criteria are best thought of as *concept-schemata* rather than as concepts. The point is that there *need* be nothing in common to *all* the members of the family: the criteria for applying the term may at one time (for one family member) be A, B, and C; for a subsequent use, B, C, and D; and for a still later one, D, E, and F. Yet those distinct sets may nevertheless be related by the fact that there were reasons, at a particular point, for rejecting A and replacing it with D, and so forth, eventuating by a rational process in the criteria D, E, and F. And if there happen to be criteria in common, even through the entire lifetime of the concept-schema, those common criteria have no special status: they are in principle as subject to revision as any others. They just happen, as a matter of contingent fact, not to have been revised. [Shapere (Unpublished b)] One of the most fundamental errors of traditional philosophy has been to suppose (demand?) that there must be criteria in common; this is what I have called [Shapere (1984), Ch. 11] the 'Platonic fallacy'. (Philosophy is indeed a footnote to Plato – alas!)

In a review of N. Nersessian's *Faraday to Einstein: Constructing Meaning in Scientific Theories*, Patrick Enfield says, 'Nersessian seeks to incorporate Shapere's suggestion that the reasons scientists give for changing meanings of terms provide continuity of meaning and hence commensurability. The

34

claim that these reasons determine meanings is characteristically ambiguous, although textual evidence indicates that it is not the trivial idea that different reasons tend to lead to different concepts. It appears rather to be the view that the chain-of-reasoning is actually a part of the resulting concept – hence the necessity, and not just desirability, of the detailed historical investigation. It would seem to follow that two people could not arrive at the same concept by different routes.' [Enfield (1985), pp. 641–642]. I trust that the alleged but unspecified ambiguity has been removed by my disussion: the chain-of-reasoning is 'part of the resulting concept' not in any essentialist sense, but only in the sense that what is important in trying to understand scientific concepts is, as I have tried to show, the way they develop; and to this task traditional theories of meaning, and indeed the whole program of studying meanings, is a hindrance rather than a help. For instance, only if one holds (as neither Nersessian nor I does) that the chain-of-reasoning is essential to the meaning will it follow that two people could not arrive at the same concept by different routes, so that two distinct chains-of-reasoning give different meanings. My view obviously makes it possible for there to be further reasons for saying that two otherwise distinct chains of reasons lead to, or have to do with, the same concept.

21. Shapere (1984), Ch. 18.

An example of central importance for the philosophy of science, but which is ignored by modern linguistic theories of reference, is this: Do idealizations, like 'mass-point', refer? That question is not answerable in terms of a general linguistic (i.e., scientific content-free) analysis of the concept of reference: what counts as an idealization is determinable only in contrast to concepts which are *not* conceptual devices; and that contrast is made, in science, on the basis of internal scientific reasons. (More accurately, it has become an ideal of science to be able to make the distinction solely on the basis of internal scientific considerations, the distinction having earlier been made on far looser and more general bases. But as a matter of contingent fact, science has been able to make the distinction, in a large number of cases, on the basis of internalized reasons.) Indeed, the distinction cannot be made in the simplistic yes-or-no terms which philosophical analyses imply, since idealizations are parasitic on what I have called 'existence-concepts'. [Shapere (1984), Ch. 17, and in unpublished writings on the more general subject of what I have called 'conceptual devices', a term which covers a variety of types of concepts or usages often distinguished by such names as 'idealizations', 'abstractions', 'models', 'fictions', 'simplifications', etc.] Nowhere is the idea of internalization better illustrated than in the process by which decisions as to whether and in what sense a concept 'refers', and what it is to refer, than in the internalization of the criteria for saying that a concept (or proposition) is a conceptual device, and of the sense of saying that a scientific concept 'refers'. Whereas once upon a time people supposed that 'All bodies must be extended' was a synthetic *a priori* truth, unassailable by any scientific results, and that that truth determined that treating bodies as 'mass-points' was an idealization, we have learned how to give *reasons* for saying that it is such. Those reasons consist of background beliefs which have shown themselves successful and doubt-free, so that whether a concept is an idealization (or, more generally, a conceptual device) is a contingent matter, to be determined in terms of what we have found reason to believe. For a study of the varying status of unextended bodies in the history of science, see Shapere (1984), Ch. 17.

22. Shapere (Unpublished a), (Forthcoming).
23. Shapere (1982).
24. *Cf.*, Shapere (Forthcoming).
25. Detailed arguments for this view are found in Shapere (1986b) and (Forthcoming). See also Part III, below.
26. See Shapere (1982), (Unpublished a).
27. In discussing Shapere (1982), Hacking declares that 'Shapere had a more philosophical purpose in his analysis of observing. He holds that the old foundationalist view of knowledge was on the right track. Knowledge is in the end founded upon observation'. [Hacking (1983), p. 185] Hacking realizes that I do not agree with the foundationalist view that observation-claims are incorrigible: '[Shapere] notes that what counts as observations depends upon our theories of the world and of special effects, so that

there is no such thing as an absolute basic or observational sentence'. [Ibid.] However, the point requires further emphasis. 'The given' must be arrived at by a very complex process of investigation which includes bringing to bear a great deal of background belief which is presumed to be successful, doubt-free, and relevant. In any of these respects error is possible, as is the result arrived at and taken to be 'given'. Nevertheless, it is also possible to remove all reasonable doubt from acceptance of that given, and in that case, when combined with the well-grounded (successful and doubt-free) beliefs which constitute the relevant background information of the experiment or observation, that given can serve as rational ground for accepting, rejecting, or modifying the hypotheses under investigation, or for leading us to new hypotheses, or even for rejecting some of the background information itself. Thus, according to my view as Hacking puts it, 'the fact that observing depends upon theories has none of the anti-rational consequences that have sometimes been inferred from the thesis that all observation is theory-loaded'. [Ibid.] Why Hacking then concludes this discussion with the following remark is wholly inexplicable: 'Thus although Shapere has written the best extended study of observation in recent times, in the end he has an axe to grind, concerning the foundations for, and rationality of, theoretical belief'. [Hacking (1983), p. 185] This makes it sound as if my opposition to irrationalism were simply an unquestioned, unargued dogma. But on the contrary, that opposition rests on very detailed and careful arguments and analysis. For instance, Shapere (1984), Part I, raises a host of objections against a variety of forms of irrationalist views of science. And in (1986b), supplemented by other recent works, a thorough analysis, resting on detailed arguments, is provided for why certain kinds of considerations rather than others are employed in particular scientific problem-situations, and furthermore for why such considerations count as reasons. In what sense, then, can those conclusions be said in fairness to consist in grinding an axe?

Another of Hacking's claims is equally mystifying. He says that 'On this view, something counts as observing rather than inferring when it satisfies Shapere's minimal criteria, and when the bundle of theories upon which it relies are not intertwined with the facts about the subject matter under investigation. The following chapter, on microscopes, confirms the force of this suggestion. I do not think that the issue is of much importance'. Hacking gives no argument for the latter claim. But surely, for understanding the scientific enterprise, one of the most essential and important tasks is to gain insight into how we decide rationally that the results of observation and experiment can be used for testing theories or arriving at new ones. It is certainly true, as Hacking says, that 'Observation, in the philosophers' sense of producing and recording data, is only one aspect to experimental work' (Ibid.); but that fact does not reduce its importance. Hacking implies that another aspect of that work is of greater importance: 'It is in another sense that the experimenter must be observant – sensitive and alert'. (Ibid.) Certainly that is true (though that does not necessarily mean that he must watch the experiment carefully – see Shapere [1982], pp. 508 ff.). But it is what he must be alert *for* that imposes the requirement that he be alert, and that is what is of central importance.

Incidentally, the statement that 'something counts as observing rather than inferring . . . when the bundle of theories upon which it relies are not intertwined with the facts about the subject matter under investigation' is not correct. In the solar neutrino case, weak interaction theory is intertwined with the theory of the source (the subject matter under investigation) as well as the theory of the observing instrument; yet the objectivity of the inquiry, and its counting as an observation-situation, is not vitiated by that fact. [*Cf.* Shapere (1982), especially pp. 516 ff. The final sentence of the footnote on p. 516, and the general question of how objectivity is preserved in such (ubiquitous) situations is further clarified in Shapere (Unpublished a).]

28. Shapere (1986a), (1986b), (Forthcoming), (1984).
29. See Shapere (1984), Ch. 15.
30. Our everyday experiences of learning acquaint us with the experience of recognizing what *can* be achieved, and learning, from that recognition, to change our goals accordingly. Methods that cannot be applied, goals that cannot be achieved, are altered when we discover their limitations. Sometimes, too, revision of our views of nature can show a method hitherto thought inapplicable, or a goal

hitherto considered unachievable, or a problem hitherto believed unanswerable or 'unscientific', to be applicable, achievable, or answerable after all. (Consider the problem of the origin of life.)

As a first approximation, we might think of the new findings as having led to a more 'realistic' appraisal and formulation of goals, methods, problems, etc. That idea of 'realistic' must of course itself be given a deeper formulation, but how to do so is straightforward on the view outlined here: the new findings raise specific doubts about the prior conceptions.

31. It must be understood that not all beliefs which are used as background information fulfill these criteria of success and freedom from doubt to the highest possible degree. Where necessary, science relies on such less well-certified beliefs. But we must not make the mistake of saying that no beliefs fulfill the criteria: on the contrary, very many do.

The five results discussed in the present portion of this paper have not been arrived at by a process of straightforward and uninterrupted linear progress. Rather, the process has been punctuated by backslidings, errors, and misunderstandings. Furthermore, even in cases of the replacement of older theories by radically new and successful ones, often many problems are reopened that had reputedly been solved by the now-rejected theories. And finally, in such cases, often a multitude of new problems, not conceived under earlier theories, are created. Nevertheless, the long path, despite its meandering character, has in fact led to an accumulation of accepted and acceptable beliefs. *Cf.,* Shapere (1984), Ch. 9.

32. This question is dealt with in detail in (1986b) and (Forthcoming). It and most of the questions that follow are discussed to at least some extent in the remainder of the present paper.

33. Shapere (1986b), (Forthcoming).

34. Shapere (1986b), (Forthcoming), (Unpublished b). The subjects or areas listed here as possible competitors of science are ones which have been brought up by questioners in discussions following my presentations of papers on a number of occasions. I am particularly indebted to Steven Stich, however, for an especially thoughtful and thorough letter regarding this question.

35. Shapere (Unpublished b), (Forthcoming).

36. Shapere (1984), Ch. 17; (Unpublished b).

37. Shapere (1984), Ch. 10; (1986b); (Unpublished b); (Forthcoming).

38. Shapere (1984), Ch. 11.

39. Shapere (1974); (1984), Ch. 15.

40. See Shapere (1984), Chs. 13–15, and, for a more up-to-date view, (1986b).

41. Shapere (1986b).

42. Shapere (1982), (Unpublished a), (Forthcoming).

43. *Cf.,* Shapere (1986a).

44. The impact of new beliefs, particularly those resulting from scientific investigation, on ethical precepts, for example, has been almost completely ignored by philosophers. Again, the reason lies in a philosophical dogma: the total independence, the mutual exclusiveness, of the 'ought' from the 'is', of the normative (the exclusive domain of ethics) from the descriptive (of which sort the scientific is alleged to be).

45. Shapere (1982), (Unpublished a).

46. Shapere (1982).

47. Shapere (1986b).

48. Shapere (1986b) describes how the piecemeal approach to inquiry led to a specification of the ideas (concept-schemata) of 'success' and 'freedom from doubt' through the piecemeal approach to inquiry. Very roughly, the discussion concludes as follows. Since the piecemeal approach isolates distinct 'domains' of inquiry for study, 'success' consists in how well a theoretical account deals with the domain items. (What constitutes a 'theoretical account of' and 'dealing with' domain items is also discussed there and in other recent work.) 'Freedom from doubt' is specified in terms of accounts of other domains, where there is reason to believe or suspect that an account of the domain under investigation should be given in the same terms as, or be compatible with, the type of account given of those other domains.

The present view most emphatically does not imply that the Greeks, say, or Kepler, were not rational. Their notions (criteria) of rationality were, however, quite different from that evolving from the piecemeal approach to inquiry; with the Greeks, for example, purely logical considerations combined with common-sense presuppositions as 'reasons'. But in spite of these differences, the later scientific idea of what counts as a reason is a rational descendant of many of those Greek ideas. [*cf.* Shapere (1986b). For Kepler, see Shapere (1984), Ch. 9, pp. 179–180.]

49. The difference between the characteristics of its domain and the domains of science provides a partial explanation of why we do not ordinarily consider Shakespeare scholarship to be a 'science' – although the difference between the two sorts of activities is far less on the present view than it has been alleged to be on previous views of science.

   Is mathematics a science or not? On the one hand, its domains are in significant ways different from those of science; but on the other hand, much of mathematics is intimately relevant to science. A brief answer is that mathematics develops either by formalizing certain aspects of science (its geometrical or topological concerns; its concern with continuous and, generally, differentiable processes; its group structure; and so on), or by departing in multifarious ways (replacing axioms, generalizing definitions and methods, etc.) from those interests to develop on its own from that point, independently of science. But despite that independent development, the results of mathematics remain either actually or potentially relevant to science: they have been found to be actually relevant in certain respects (geometry, topology, group theory, etc.), and so the departures from those respects continue to be potentially relevant.

50. As to the remaining Function 2-type questions, the fourth in the list, concerning the relations between 'reasons' and 'truth' is discussed in Shapere (Unpublished b), the fifth, having to do with the roots of the scientific enterprise, in Shapere (1984), at the end of Ch. 17, and the sixth, regarding the rise of normative principles in science, in Shapere (1984); see the index under 'normative'.

51. Quine (1969).

52. Quine (1973), p. 2.

53. Quine (1969), p. 75.

54. Ibid., p. 82. Quine's holistic view (epistemology is a component part of psychology and therefore of 'the whole of natural science' – see also ibid., p. 83) is misguided when it comes to psychology. It ignores the fact that psychology is not a part of science in the sense that, say, chemistry is a part of physics: the phenomena of psychology are *at present* very, very far from being explained in terms of the rest of science, and therefore are not an integral part of 'the whole of natural science', at least in the sense of being explanatorily interlocked with the rest of the results of science. The extent to which the science of the present (or of any other era) constitutes a 'whole' must be determined by a close examination of that science.

   Nor can we accept at face value Quine's claim that 'all this [epistemology, psychology, and the rest of science] is our own construction or projection from stimulations like those we were meting out to our epistemological subject' (ibid., p. 83) ('certain patterns of irradiation in assorted frequencies, for instance' [ibid.]). The construction of scientific (and epistemological and methodological) views involves far more than can be literally conceived as construction or projection from stimulations, as is shown in detail in Shapere (1982).

55. Quine (1953), p. 44.

56. Quine (1977), p. 229.

57. Quine (1981), pp. 24–25.

58. Quine (1969), p. 82.

59. A full and detailed account of the background information entering into this case, and of the exact roles played by that background, is found in Shapere (1982).

60. The argument can be carried one step further still, as I have carried it in (1986b): what leads us to count successful, doubt-free, and relevant beliefs as reasons is also independent of psychological considerations. We might imagine some hypothetical 'epistemological state of nature' in which an

38

ancestor of our idea of '(a) reason' stemmed from a distinction of self from non-self, and from the unpredictability or interest posed by that which was considered non-self. But that source was definitely ancestral at best; the concept of 'reason' has by now become entwined with the sophisticated beliefs of modern science, and specifically with the science that has flowed from the adoption – and ultimate transcendence – of the piecemeal approach to inquiry, and of the precise ways in which 'success' and 'freedom from doubt' are specified in that approach and its rational descendants.

61. Quine (1969), p. 75.
62. In this sense, the old positivist dictum that the concern of the epistemologist is with relations between propositions, rather than between people and their beliefs, was on the right track, even though many of them erred in supposing that the relation could be studied independently of the substantive content of the propositions themselves.
63. Quine (1977), p. 229.
64. Quine (1977), p. 233.
65. Quine (1953), p. 43.
66. Quine (1981), p. 71.
67. I do not mean to say that there is not, in some cases and in some senses at least, something like underdetermination of theories by data; in fact, there is in certain areas. What I mean is that whether there is, and in what senses, and with respect to what sorts of ideas, must be determined by an investigation of actual science and the mathematics it employs.

   Other views of Quine's also suffer from a failure to examine substantive scientific reasoning, though not all have to do with failure to grasp major lessons of modern science. An example is his well-known doctrine that 'To be is to be a value of a variable' [Quine (1953), pp. 1–19], that 'If we subscribe to our physical theory and our mathematics ... then we thereby accept these particles [spoken of in the theory] as real'. [Quine (1977), p. 65] This doctrine misses the profoundly important distinction in science between unqualified referring terms and idealizations, convenient fictions, and other sorts of conceptual devices. [cf. Shapere (1984), Ch. 16] All these types of terms qualify as ones that, logically, are 'quantified over', but their distinctive roles, and the reasons for treating them or their referents as the scientist does, are to be found in the background information, and not merely in the fact that the theory talks about them or that they are (can be viewed as) values of quantificational variables.

   Quine's approach to his problems is deeply conservative. His interests in the Duhemian problem, for instance, focus on how we can *preserve* any particular belief, not on how we might decide to accept a new one. Again, scientific change is said to be constrained by 'our natural tendency to disturb the total system as little as possible'. [Quine (1953), p. 44] From an epistemic perspective concerned with scientific change, however, the really important issue is not how we manage to maintain as much of our older views or preferences as we can, but rather how we manage to decide on new directions and new beliefs.

   For further comments on Quine, and in particular on his antirealist view of science, see Shapere (1984), pp. 254–256.
68. Rorty (1979).
69. According to Rorty, what must be rejected is the idea that knowledge consists in correspondence to reality. However, in Shapere (Unpublished b), it is shown that something important – having a functioning role – remains in science – is a rational descendant – of the correspondence theory of truth after all.

# Bibliography

Bacon, F. (1939), *Novum Organum*, selections reprinted in *The English Philosophers from Bacon to Mill*, E.A. Burtt, ed. (New York: Modern Library).

Enfield, P. (1985), review of N. Nersessian, *Faraday to Einstein: Constructing Meaning in Scientific Theories, Philosophy of Science* 52: 641–642.

Hacking, I. (1983), *Representing and Intervening* (Cambridge: Cambridge University Press).

Nickles, T. (1985), review of D. Shapere, *Reason and the Search for Knowledge, Philosophy of Science*, 52: 310–312.

Quine, W.V. (1953), *From a Logical Point of View* (Cambridge: Harvard University Press).

Quine, W.V. (1969), *Ontological Relativity and Other Essays* (New York: Columbia University Press).

Quine, W.V. (1973), *The Roots of Reference* (La Salle, Ill.: Open Court).

Quine, W.V. (1977), *The Ways of Paradox and Other Essays* (Cambridge: Harvard).

Quine, W.V. (1981), *Theories and Things* (Cambridge, Harvard).

Rorty, R. (1979), *Philosophy and the Mirror of Nature* (Princeton: Princeton University Press).

Shapere, D. (1986a), 'External and Internal Factors in the Development of Science', *Science and Technology Studies,* Vol. I, No. 1 (includes replies to commentators).

Shapere, D. (1986b), 'Objectivity, Rationality, and Scientific Change', in *PSA 1984,* Vol. II, P. Kitcher and P. Asquith, eds.

Shapere, D. (Forthcoming), *Observation and Knowledge* (Oxford: Oxford University Press) (Formerly titled *The Concept of Observation in Science and Philosophy).*

Shapere, D. (Unpublished a), 'Observation, Experiment, and Interpretation'.

Shapere, D. (1974), 'On the Relations Between Compositional and Evolutionary Theories', *Studies in the Philosophy of Biology,* F. Ayala and T. Dobzhansky, eds. (London: MacMillan): 187–202.

Shapere, D. (1984), *Reason and the Search for Knowledge* (Dordrecht: Reidel).

Shapere, D. (Unpublished b), 'Reason and the Search for Truth'.

Shapere, D. (1982), 'The Concept of Observation in Science and Philosophy', *Philosophy of Science,* 49: 485–525.

# 'Twixt Method and Madness[1]

THOMAS NICKLES

University of Nevada, Reno

## 1. Introduction

For years I have urged that there is more to scientific method than meets the philosophical eye. Some highly touted conceptions of method – the hypothetical-deductive (H-D) method, for instance – are remarkably thin when it gets down to details. For when one studies real scientific cases of problem solving and theory construction, the methodological features are almost always more substantive and more interesting than what one reads in general methodological accounts. 'There are more things in heaven and earth . . .'

As thanks for my efforts, I am now asked to write on the methods that I and other philosophers of science use (or should use) in thinking about scientists and *their* methods. This second-order task may be more difficult than the first-order task of writing about science itself. In recent history at least, philosophy has become the domain of problems that have proven intractable to 'scientific' treatment – a loose collection of subject matters for which no special methods or techniques are known to succeed. What then can I say about *philosophical* method? I limit myself to four preliminary remarks.

First, I am guilty of exaggeration. Philosophy is not devoid of modes of inquiry and of powerful intellectual tools. But neither is it limited to formal, 'purely analytical' tools. I consider philosophy of science a broadly empirical subject, although I do not view methodology as a narrow, rigorous 'science of science'. Methodology is certainly *scientific* and involves the application of scientific methods to the evaluation of scientific methods themselves, but methodology is not a *science,* for science is not itself a homogeneous domain which could be the subject of another, rigorous science. I join those who seek a more naturalistic and historical-sociological account of science and depart from those who see philosophical method as ultimately *a priori.* This does not mean that I abandon the formal analysis of scientific and methodological ideas, although I do conceive all justification as ultimately social. Narrowly defined, naturalism becomes scientism, an enemy of historicism. I define it more broadly.

Second, we must not assume that correct philosophical, historical, sociological, or

*Nancy J. Nersessian (editor), The Process of Science. ISBN 90-247-3425-8.*

© *1987, Martinus Nijhoff Publishers (Kluwer Academic Publishers), Dordrecht. Printed in the Netherlands.*

psychological methods will be identical with the methods of the branches of science they study. For example, it is perfectly possible for an historian to discover inductively that scientific practice is H-D. Equally, we might discover by H-D methods that scienctific research is inductive.[2] In fact, we discover by *all* of these methods that science employs *none* of them exclusively. There is not just one correct method which is best for all fields of inquiry at all times.

Third, we must not even assume that there is one big, scientific method common to all *sciences* or to all subspecialties of the *same* science. There is not just one thing called 'science', defined by one, special method. But neither are the scientific disciplines methodologically disjoint.

Finally, that there is no Main Method or Rigid Plan does not imply that there exist no methodologically interesting routines by which specialists address important classes of problems. Philosophers torment themselves by first demanding that genuine methods be routine procedures and then proceeding to ignore the many routine problem-solving methods that do exist – as too specialized and therefore not philosophically (or methodologically) interesting. Such philosophers make impossible demands on methodology. Seventeenth-century methodologists such as Bacon, Descartes, and Newton would have been delighted to have the mathematical, experimental, and theoretical problem-solving techniques available today. They would have considered such devices genuine methods and their study a proper part of methodology.

By *method* I mean the procedures which scientists themselves use to achieve their goals. Usually, the goals are solved problems, whence the methods provide some guidance to searching for and testing problem solutions. As hinted above, I by no means restrict the term 'method' to algorithms and sausage grinders. By *methodology* I mean the study and comparative evaluation of methods. Taking a page from Hans Reichenbach, we may say that methodology divides into descriptive, critical, and advisory tasks.[3] I cannot adhere faithfully to his understanding of these tasks, but even Reichenbach gave sociology and psychology of science a role, particularly in carrying out the descriptive task. Methodology is not the special domain of philosophers, few of whom are competent to provide scientific advice. But neither does methodology exhaust philosophy of science.

Not all methods are equally good. Some methods aspire to achieve unrealizable goals, and some methods are poor ways of achieving attainable goals. Since methods govern research processes, methods can be evaluated along three primary dimensions: the suitability of research *goals,* the efficiency of the *processes* in producing research products (e.g., problem solutions), and the quality of these *products.*

Most philosophers have been content to discuss the quality of the products, which dimension resolves into two others: epistemic quality (truth, probability, degree of corroboration, reliability) and 'interest' or 'content'.[4] But economy of research is just as important to methodology as truth and its surrogates.[5] And too many philosophers of science have been quick to announce truth as the obvious goal of inquiry and have proceeded to supply hopelessly inefficient methods of finding it. Indeed, if truth is the immediate goal of inquiry, no method of achieving it may be sufficiently efficient. To be sure, not all useful ideals need be realizable in practice; nonetheless, this fact does not justify totally ignoring practical concerns. Methodology, as the proper adjustment of means and ends to one another, is (or should be) an eminently practical subject.

## 2. Autobiographical remark

I am by nature open to alternative interpretations of scientific work and sympathetic to foreign (including nonphilosophical) points of view. Xenophobia is not one of my vices, although some will say that my territoriality instinct is underendowed. I think I have learned the lesson which history, sociology, and psychology of science should have taught philosophers by now: philosophy of science is not self-sufficient. Viable philosophical accounts depend heavily on good empirical information and even on the theoretical perspectives of other disciplines. But beyond this, I find myself actively searching for golden insights in positions widely held in disrepute. For example, it seems to me that the 17th-century discovery programs of Bacon, Newton, *et alia,* were not so logically flawed as later philosophical ridicule would suggest. I am now writing a book in which I attempt to salvage two central components of this program – the idea that methodology should be 'processual', even constructive, and the corresponding idea that conclusions obtained by constructive methods have (what I term) 'generative support', a form of epistemic support different from consequential support.[6] However, I employ these ideas in a 'local', domain-specific way rather than as methodological panaceas.

Doubtless, it was sympathy to other points of view which originally led me to take seriously what scientists themselves say in their notebooks, letters to other scientists, and scientific publications. In my early thinking about intertheoretic reduction, for example, I noticed that scientists frequently speak of the reduction of the later theory *to* a predecessor theory (e.g., the reduction of the relativistic equations of motion to Newtonian equations, in the limit of low velocities), whereas philosophical discussions of reduction almost unfailingly spoke of the earlier theory as reducing to (reduced by) the later one. These accounts equated reduction with the deductive *explanation* of the earlier theory by the later, or the less comprehensive by the more comprehensive theory, and thus missed the heuristic and justificatory functions of reduction. In fact there are several distinct concepts of reduction in science, mathematics, and philosophy.[7]

The aforementioned empirical information and the (decreasingly) radical perspectives of 'new wave' sociologists and psychologists of science not only aid philosophy in its descriptive task but also inform normative philosophy of science. Although no amount of descriptive information can produce an 'ought', it is now widely recognized that, since 'ought' implies 'can', 'cannot' implies 'not ought'. Since many well known philosophical prescriptions are unworkable when applied to real, human scientists doing real, scientific research, direct acquaintance with science and attention to historical, sociological, and psychological accounts of it can help make methodology as practiced by philosophers more relevant to the subject nominally discussed.

Nonetheless, it would be naive to take at face value everything that scientists say, particularly when they address nonscientists or interviewers from fields other than their own or when they reflect retrospectively on the development of their discipline.[8] But neither is it sufficient to examine their deeds in addition to their words. Many philosophers have examined scientific deeds, but from the wrong perspective so far as the *process* of science is concerned. If we assume that competent working scientists often employ good methodological strategies in their work and thus at least implicitly know more about method than armchair philosophers like me, it is important at least sometimes to look at scientific work *from the working scientists' own, prospective viewpoint*. This most important benefit of direct acquaintance with scientific work usually is overlooked: direct knowledge of scientific work helps one to take the working scientists', forward-looking point of view.

Sociological studies of laboratory life take a large step in this direction, although they so far fall well short of full participant-observation, and they look mainly at laboratory research alone.[9] A genuinely historical study of an ongoing piece of research and, perhaps surprisingly, even artificial intelligence (AI) can provide other ways of assuming the point of view of a working scientist.[10] Once we adopt this viewpoint, important changes in our philosophical emphasis are inevitable. Obviously, there is not one single 'working scientists' viewpoint', but this fiction is more useful than harmful for what I have to say here.

In what follows, I shall indicate how I try to integrate philosophical, historical, sociological, and formal studies into coherent visions of the various sciences. My emphasis here will be on the working scientists' point of view rather than on the historicity of science. Since I touch on too many topics to deal with each one, I frequently refer the interested reader to my published and forthcoming work for details.

## 3. From retrospect to prospect

Point of view can be important in problem solving. One need think only of the arguments of Buridan, Oresme, Copernicus, and Galileo that, since all perceived motion is relative, experience alone does not decide the question whether it is the earth or the heavens that move. The young Einstein imagined himself moving with the speed of a light wave and thereby won a great insight into the propagation of light. More recently, in a sociobiological dispute, Richard Dawkins maintained against his opponents that the term 'fitness' tempts us to adopt the biological individual's point of view – either that of the parent or that of the offspring – whereas 'There is really only one entity whose point of view matters in evolution, and that entity is the selfish gene'.[11] Dawkins's own (controversial) account was written from this novel point of view. In a more literal sense, many psychologists and social scientists have found it fruitful to adopt the agent's viewpoint, on some occasions at least. There is evidence that point of view may be a factor even in solving puzzles such as the missionaries and cannibals problem.[12]

In the case of methodology of science, I contend that adopting the working scientists' point of view not only aids in solving methodological problems but also helps to determine what the methodological problems are in the first place. Not all the methodological problems that exercise philosophers correspond to actual problems of working science; and certainly not all scientifically interesting methodological problems are sufficiently noticed by philosophers. What follows is an overlapping but non-exhaustive list of methodological benefits that accrue from adopting the scientists' own point of view. I shall have space to expand on a very few of these themes in subsequent sections.

*a. From product to process.* While it certainly is important to study the logical structure of the products of scientific investigation and the degree to which those products are justified, to confine philosophy to a static or retrospective analysis is to ignore half of scientific methodology.

*b. From solutions to problems; from theory choice to problem choice.* In particular, philosophers have worried a great deal about what theories are, about theory choice, and about the empirical underdetermination of theories. They have devoted little attention to what problems are, where they come from (how they arise), and how scientists generate or choose their problems, projects, and research programs. These choices, too, are underdetermined. There can be no question that these issues are fundamental to methodology. For example, that it generates interesting new problems can be a central reason for choosing one research program over another. Producing problems can be an important scientific *achievement*, with a major role in one significant form of scientific justification.[13]

*c. From truth to the solved problem.* Traditionally, philosophers have been interested in absolute truth and in seeing things *sub specie aeternitatus*. However, the history of many branches of physics indicates that reliability is all that is required for dramatic progress in research. While questions about truth are terribly important, truth has been overrated as a methodological topic. Overconcern with truth has damaged the project for a process methodology by imposing a static, timeless (or else, retrospective) stance that precludes adopting the scientists' view of the research process.

Most scientific research is problem-solving activity. If philosophers discuss problems at all, the problems frequently are considered incidental obstacles to the attainment of truth. Since the general truth about the world cannot be known in any absolute sense, philosophers have difficulty stating in what sense science achieves anything or progresses. But if we follow Thomas Kuhn in taking the solved problem (research puzzle) rather than the established truth (e.g., an established-true lawlike claim) as the unit of achievement, these difficulties are less serious.[14] The central problem of methodology is surely the underdetermination of theory, a philosophical conundrum on which philosophers spin their wheels. Yet working scientists manage to deal with this difficulty – primarily by refusing to divorce scientific problems from the richly textured constraint contexts in which they arise. The research process continues without having to settle questions of ultimate truth. Scientists work at a moving front, where this 'motion' is relative to previously 'established' results.

'Established results' are reliable results, but they need not be true results, much less results *known* to be true. Truth seems more important to some scientific disciplines than others. In all disciplines we can state some (sufficiently qualified) claims which we take to be established as true. However, scientists in some fields, such as molecular genetics, are more confident that many of their high-level claims are true than scientists in other fields, such as particle physics. The former tend to require that research be based on claims which are not just 'reliable' or 'correct for the domain in question' but 'true' – and hence *completely* reliable. The latter make do with lower grades of reliability. But in no case, it seems to me, does a 'realist' methodology of science provide constraints on problem solving that are not available to the irrealist or to the agnostic about truth. Recently, realists have advanced several arguments which purport to show that realists can do all kinds of things that irrealists cannot, but I find these arguments – which I cannot discuss here – unconvincing.

But is not truth important to research?, it will be asked incredulously. Of course, in a trivial sort of way. Researchers must not falsify data, conceal theoretical weakness, etc. My point is different. Adopting the solved problem and the reliable result, rather than the certifiably true result, as the immediate goal of inquiry is not to throw away something that we ever had. Since epistemological foundationism fails, we have no 'direct access' to the truth about the world. All access to the truth is mediated by inquiry procedures available to all suitably trained investigators. Requiring that they be true introduces no additional, practical constraint on problem solutions (theories, models, explanations) beyond the usual requirements of empirical support and consonance with previously established results. Since this is so, I think that a process methodology should be a problem-solving rather than a directly-truth-seeking account of scientific research. In making this recommendation, I myself am not abandoning truth for an irrealist view of science; rather, I am 'bracketing' the issue of truth. If some solved problems can reasonably be construed as true claims about the world, so much the better. We certainly hope that some of our more reliable results are true. But knowing their truth status is not necessary for research to proceed.

*d. From logic to heuristics and from justification to discovery – and back again.* All phases of research, including 'final justification' as written up in the proverbial 'final research report', involve problem solving. Yet simple logical relations from theory to available data do not take us very far in understanding scientific problem solving – which partly explains why the positivists, Popper, and company consigned problem solving, under the rubric of 'discovery', to psychology and anecdotal history. For Popper there is little to say about great scientific discoveries other than that they are the result of an almost divine inspiration.

Paul Feyerabend's attack on method is well known, but before him, Popper denied that there is any such thing as scientific method.[15] Of course, by 'method' Popper had in mind a general, fail-safe algorithm for generating good theories. He was correct that there is no scientific method in that sense. But it hardly follows that there is nothing more than inspired hypothesizing followed by severe testing. Work in artificial intelligence has done much to show that heuristics is a legitimate methodological concept and more – a model for methodology. Heuristics falls between anecdotal psychology of discovery and algorithms. Heuristics can be precise enough to be logically respectable, yet fallible enough that there is no question of reducing method to fail-safe recipes. Further, it is the business of heuristics to address problems of economy of research – the central problem of process methodology, broadly speaking. Heuristics is no single, fixed doctrine; rather, it consists of growing and changing families of (often content-specific) search procedures which can be adapted to the special problems at hand. Heuristics falls between method in the sense of algorithmic rules and the divine madness of Popper's 'inspiration', between Plato and the poets, as it were. Hence my title: 'Twixt Method and Madness'.

Ironically, the neglect of heuristic discovery methods by philosophers has meant the neglect of much theory of *justification,* for heuristic appraisal is one of the most important forms of justification in science. Almost all confirmation theories developed by philosophers are retrospective. A theory is judged entirely by its empirical record of past success and failure. By contrast, the motto of working scientists is, in the immortal words of John Dewey, 'We live forward'. By its very nature, heuristic appraisal is forward looking rather than backward looking. Heuristic appraisal is the evaluation of the future prospects of a theory, hypothesis, research proposal, or program, its *generative potential,* i.e., its apparent ability to handle problems still outstanding and to generate interesting new questions for research. The urgent question which working scientists have for a given theory or program which they are actively using or considering is not 'What have you done for me?' or even 'What have you done for me lately?' but rather, 'What can you do for me *tomorrow?'*

The working scientists' point of view is almost never entirely retrospective. To be sure, scientists do wish to clear up blemishes in theories which already constitute fairly reliable bases for research. However, the primary motivation for this is to improve the general security of using these older results and/or to gain insight into a particular puzzle of current research interest. Roundly stated, scientists look back only with an oportunistic eye out for hints of future developments – as when they model attempted solutions to new problems on successful old solutions, as rationally reconstructed. Even the 'final research report' justifies the importance of the findings in part by reference to their future applicability to new problems; in this sense, even 'final' justification has a heuristic dimension. Only recently have a few philosophers such as Urbach, Wimsatt, and Nickles discussed heuristics as a form of justification.[16] Scientists are not nearly as interested in reconfirming claims as in seizing opportunities to do something new – something that will bring them recognition. For example, standard texts and chapters on methodology attach great importance to exact replication of results as central to scientific objectivity, yet this sort of replication almost never occurs, unless as a didactic exercise.[17]

Discovery issues are also important to 'final' justification in a more epistemic way, which I term 'justification as potential discovery' or 'discoverability'. The basic idea is that a theory is justified by seeing how to derive it usually in a highly reconstructed way from what we already 'know' about the world. The associated concept of generative support is quite different from, but complementary to, the familiar idea of consequential (predictive) support.[18]

*e. From unlimited resources to economy of research.* Taking the working scientist's point of view also requires serious attention to economy of research. While 'brute force' exhaustion methods are sometimes available, and while good fortune very occasionally rewards sheer guessing, methodology in a traditional sense going back to Socrates was related to *technē* and fell between rigid, mechanical procedures requiring no art or judgment and inefficient procedures which succeeded (if at all) by chance or accident. (See Plato's *Phaedrus* 265D–277C and Hippocratic medical texts.) There is a long tradition relating method to efficient procedures. Today even algorithms are judged in terms of their efficiency and are sometimes replaced in research by heuristics, which may be more efficient although risky. Economy may outweigh security.

Until recently, philosophers concerned themselves little with the practicalities of research economics, with disastrous results. For example, given the immense size of the domain of possible theories to be considered in any area of inquiry, Popper's method of more or less blind conjecture plus attempted refutation has virtually no chance of finding the correct theory by eliminating all the others – no matter how long the method is applied – even though Popper's method is self-corrective.

While philosophers traditionally have been optimizers or maximizers, Herbert Simon has shown how naive this policy can be. In some cases, optimizing is rational only if infinite time and/or resources are available. (An extreme case is someone who insists that we have the absolute truth of the matter at hand before we proceed to the next step.) Rather, we must make do with satisficing. Our rationality is 'bounded'; time and resources are limited.[19]

If rationality is a matter of deliberation and experienced judgment as a member of a community of peers rather than the application of formal rules by isolated individuals, then our rationality is additionally restricted by the fact that we cannot be expert members of more than a few communities.[20] About most things we can be only semi-rational – e.g., well enough informed to know who the reliable authorities are. To some extent our rationality is founded by what we know. And since few philosophers are members of the scientific communities whose fields they discuss ...! These pessimistic, fallibilistic thoughts are confirmed by psychological research.[21]

*f. From global to local.* By paying close attention to the words and deeds of scientists engaged in unfinished research, we can better appreciate that the more interesting methods are highly discipline- and context-specific. By contrast, excessive generality has been an Idol of the Tribe of Philosophers almost from the beginning. Conceiving method as a content-neutral 'logic' of science makes methodology equally applicable to all possible worlds – and correspondingly weak or useless. Even methods which assume a certain degree of regularity in our universe are thin insofar as they aspire to be general methods for any science whatever.

In our era, the positivists and Popper were keen to establish the methodological unity of science, for understandable historical reasons. But given the great (and well deserved) influence of these writers, the (perhaps unintended) result was an impoverishment of process methodology. If one looks only for methods that all sciences have in common – the 'least common denominator', so to speak – one necessarily abstracts away from the rich and sometimes powerful methods available for specific kinds of problems in the special fields. For example, there is only so much that can be said about inductivism-in-general versus hypotheticalism-in-general, and most of that was said long ago.[22] Actually, the movement from inductivism and other 'constructive' methodologies to hypotheticalism and 'critical' methodology is at once a move from *process methodology* to *product methodology,* or, very roughly, from the kind of methodology practiced by early British methodologists such as Bacon and Newton to that practiced by German methodologists, such as Kant. I should like to see this movement reversed, though not for a return to classical inductivism.

Methodological claims that fail to be globally applicable or which fail to hold for the gigantic theory changes usually discussed by philosophers (Relativity Theory and Quantum Theory vs. Newtonian Mechanics, etc.) may nevertheless hold for some sciences and/or for local research contexts within them. Thus Ernest Nagel's account of intertheoretic reduction[23] fails for the supertheories, but it nevertheless succeeds for some smaller theories. The twin suppositions that methodology belongs to philosophy exclusively and that philosophical concerns are completely general (pertaining to the 'join' of all sciences) have tempted philosophers to embrace a pattern of unsound methodological inferences.[24]

A further point is that, given the immense variety of scientific problems, completely general methodologies of the research process are likely to be weak. Writes E.A. Feigenbaum:

There is a kind of 'law of nature' operating that relates problem solving generality (breadth of applicability) inversely to power (solution successes, efficiency, etc.) and power directly to specificity (task-specific information).[25]

This is the distinction between general problem-solving methods and 'knowledge-based' or 'expert' systems. In respect of power and sensitivity to specific information, the H-D method is not much of an advance on inductivism.

*g. From anti-circularity to bootstrap.* On the Kantian conception of philosophy as legitimator of all fields, including the sciences, philosophy could make no justificatory use of empirical information in arguing the possibility or validity of, say, physics or biology, upon pain of circularity. Such a view was commonly expressed in philosophical discussion until recently. As is well known, foundationist epistemological views induce a stratification within the body of scientific statements. Observational data is employed to test theories; hence, observational data cannot be supported by theoretical considerations, upon pain of circularity. Justification must be linear, not just to avoid logical circularity but to preserve the intrinsic epistemic order among propositions. A sophisticated version of this idea, defended by Imre Lakatos and his successors, John Worrall and Elie Zahar, is that no information used in the construction of a theory can count in its support – for that would be to use the same information twice.[26]

Unfortunately for the philosophers, scientific practice is not at all like this and yet seems quite successful – and not irrational. As Hilary Putnam long ago noted,

Justification in science does *not* proceed 'down' in the direction of observation terms. In fact, justification in science proceeds in any direction that may be handy – more observational assertions sometimes being justified with the aid of more theoretical ones, and vice versa.[27]

Exit the linear, foundational view of scientific justification.

But what do we put in its place? The answer has been some kind of 'bootstrap' view of scientific testing and justification. Clark Glymour[28] has given formally precise expression to the idea, but the intuitive notion, applied to methodology as well as to substantive scientific claims, can be found in many writers. In the final line of Book I of the *Novum Organum,* Bacon wrote: 'the art of discovery may advance as discoveries advance'. More than two and a half centuries later, Charles Peirce wrote that 'Each chief step in science has been a lesson in logic'.[29] A more radical but less elegant pronouncement comes from another pragmatist, John Dewey, in his *Logic: The Theory of Inquiry:*

All logical forms . . . arise within the operation of inquiry and are concerned with control of inquiry so that it may yield warranted assertions. This conception implies much more than that logical forms are disclosed or come to light when we reflect upon processes of inquiry that are in use ... [I]t also means that the forms *originate* in operations of inquiry ... [W]hile inquiry into inquiry is the *causa cognoscendi* of logical forms, primary inquiry is itself *causa essendi* of the forms which inquiry into inquiry discloses.[30]

Recently, Dudley Shapere has expressed a similar, bootstrap thought in terms of 'learning to learn'.[31] In the past few years, sociologists of science have made a related point in arguing that reality cannot be invoked to explain why scientists behave in the ways that they do. I shall return to these ideas briefly at the end.

*h. From 'apriori' intuition to empirical research.* Taking into account the behavior of real, flesh-and-blood scientists demands respect for empirical information about scientific work. This information is now available in rapidly increasing quantities from history of science, sociology of science, psychology of science, and direct study of contemporary work. Yet, as James Woodward insightfully remarked to me, even the best, recent, philosophical work on confirmation theory, explanation, and other topics constantly attempts to explicate our 'philosophical intuitions' about these topics. Where did these intuitions come from? Doubtless, some of them are correct. But recent developments in linguistics and philosophy of language (from positivistic and ordinary language philosophy to empirical linguistics) should have taught philosophers of science a lesson: There is no short cut to the empirical understanding of science. In the first place, given the important differences in undergraduate and graduate training and in research experience, philosophers' intuitions may be quite at variance with scientists' intuitions. But in the second place, *no one's* intuitions can substitute for tests against reliable empirical data and theories, as Peirce already observed in his famous attack on 'the *a priori method*'.[32]

At the methodological level, as Michael Friedman observes, no method is *a priori* best. 'There is no inductive method that is more reliable in every logically possible world than every other method'.[33] Given that efficient methods are empirically committed, domain-specific methods, it is an empirical question which methods work better for which kinds of problems. The methods of science themselves must be evaluated scientifically. Methodology must itself be an empirical, scientific subject.

## 4. From global to local

As Herbert Butterfield famously observed, general history, in contrast to research monographs, can hardly fail to provide a distorted picture of historical developments.[34] General history reduces the vagaries and vicissitudes of the historical process to a few simple patterns and makes it appear that all history was aiming at the present as its goal, its *terminus ad quem*. Such 'whig' history is so far from being genuine history that the term 'general history' is practically an oxymoron.

The same can be said for the term 'general methodology', in my opinion. Yet many philosophers still equate methodology with the study of inferential and developmental patterns which all sciences have in common. At once this equation places a premium on (1) demarcating science from pseudoscience, metaphysics, and nonscience generally, instead of, say, demarcating good scientific work in a particular area from bad work or differentiating work within one specialty or on one type of problem from other specialized work; (2) esablishing the unity of science – science as a more or less homogeneous field for philosophical study; (3) finding completely general principles, rules, or 'laws' for this domain.[35] For positivistic writers, these rules were largely *a priori* laws of logic. For a more recent generation of historically motivated philosophers, they are general models of scientific change. Whether logical or historical, any *general* methodology or model of science – one which aims to capture all sciences, possibly at all stages of their historical development – will be *formal* in the sense that it cannot link methodology with the specific *content* of any particular specialty at a particular time.

There are many reasons why this view of methodology came to be espoused, especially in its more logical versions. One reason was an understandable infatuation with the rigor and power of the new symbolic logic. Another was the influence of the Kantian conception of philosophy as legitimator of disciplinary goals and methods and as adjudicator of boundary disputes. A third was a fascination with the Kantian problem of distinguishing our human, *a priori* contribution from 'nature's' purely factual contribution to the body of science. Reichenbach and several other positivists were neo-Kantians, and philosophers still tend to think of their discipline in this way, even though few scientists pay any attention to them. The development of relativity theory and, to a lesser extent, quantum theory had a profound effect on the transformation of neo-Kantian ideas into modern logical positivism.[36]

Given this view of philosophy as furnishing the General Methodology of Science, there are obvious sociological and psychological reasons for its persistence, and there is a professional bias toward generality. (Naturally, alternative visions of the profession will have their own biases.) For one thing, the very possibility of methodology, on this conception, presupposes the unity of science. So it is understandable that there has been a strong bias toward unity in philosophy of science. For another, philosophers are afraid to abandon the view that methodology is general. They fear that there will be nothing for them to do if it turns out (as I think it does) that most of the still interesting methodological problems are specific to the individual scientific disciplines and subdisciplines. Philosophers fear that they will lose their self-identity and their territory. There is a great irony here, if I am right, for this very conception of philosophy and method prevents philosophers from saying much that is interesting about method. By imposing virtually incompatible demands on the subject, in order to guarantee themselves something to say, philosophers leave themselves with almost nothing to say, beyond saying that scientific methods do not exist!

Domain-specific methodology is sometimes dismissed with the quip that 'Philosophers need not bother about how to wash test tubes'. Admittedly, there are some fairly general, powerful methods that philosophers have had a hand in developing, e.g., methods of statistical inference and experimental design. Furthermore, there are topics such as explanation and confirmation which *can* be treated in a fairly general way. But scientists do not employ the same experimental design for every research problem; and philosophers are learning that even the most general accounts of explanation and confirmation are not equally applicable to all domains (a point that I argued in some early papers on explanation).[37] Recent work by psychologists of science confirms what history of science teaches, *viz.,* that one confirmation strategy or explanation strategy may work best in one world (and in one domain within that world) and quite another strategy in another world (domain). For instance, in one problem context, a confirmation strategy may be more efficient than a falsification strategy.[38]

Just as general history tends to become whig history, it seems to me that general methodology tends to become *whig methodology*. What do I mean by 'whig methodology?' First, general methodology, by reducing various kinds of research to their least common denominator, tends to oversimplify by forcing scientific work into a few basic patterns. The simple 'logic of science' which emerges is just as misleading as the simple patterns of historical development in whig history. Second, by assuring us that what really matters to research is revealed by the simple patterns disclosed by the philosopher of science, inquiry into the *process* of scientific inquiry is discouraged and the open-future viewpoint of the working scientist is missed or ignored. Scientific results appear to be the natural and expected products of the method applied to the data of nature rather than the result of a good bit of human construction and human judgement. It becomes tempting to explain research only in terms of the ways in which nature constrains our thought, when in fact this conception of nature was precisely what the scientists were in the process of hammering out. The upshot may be a view of research that is too realistic (in the sense of convergent, epistemological realism[39]) and retrospective rather than prospective. Scientists reflecting on their careers tend to fall into the whiggish-realistic way of thinking, as Andrew Pickering observes:

> Missing from the scientist's account . . . is any apparent reference to the judgments entailed in the production of scientific knowledge – judgments relating to the acceptability of experimental data as facts about natural phenomena, and judgments relating to the plausibility of theories. But this lack is only apparent. The scientist's account avoids any explicit reference to judgments by *retrospectively adjudicating upon their validity* . . . Theoretical entities like quarks, and conceptualizations of natural phenomena like the weak neutral current, are in the first instance *theoretical constructs:* they appear as terms in theories elaborated by scientists. However, scientists typically make the realist identification of these

constructs with the contents of nature, and then use this identification retrospectively to legitimate and make unproblematic existing scientific judgments.[40]

By being endowed with reality, the constructs become self-justifying, self-certifying: the assumption of their existence is employed in the very (retrospective) arguments used to justify these existence claims. This phony reasoning amounts to taking their existence as given and ignores essential features of the research process which originally led the research community to license them as valid. Pickering notes that philosophers rarely make this mistake anymore, yet strong realists face a constant temptation to think in these terms.[41]

Whig methodology is not a sin found only among positivists and classical Popperians. Surprisingly, much *historical* methodology is whig methodology. Ironically, historical methodology has transformed General Methodology into Super-General Methodology, that is, super *abstract* methodology. A little empirical knowledge can be a dangerous thing! The line of argument underlying much philosophical work in the past two decades runs something as follows:

1. (Presupposition) An adequate methodological account of X (explanation, confirmation, etc.) must be completely general.
2. A single, clear counterinstance refutes a generalization.
3. Hence, a good way to test generalizations is to search for counterexamples.
4. History provides a rich source of possible counterexamples.
5. (Practical conclusion) We should therefore test methodological claims by exploring the history of science for counterexamples (and for heuristic enlightenment as to how to correct or replace those claims which fail).
6. Hence, a single historical case of acceptable[42] scientific practice can refute a methodological account of X.
7. Conversely, an adequate (unrefuted) methodological account of X must be faithful to every single case of acceptable scientific reasoning concerning X.
8. Hence, methodological accounts which are refuted by historical counterexamples are (at best) insufficiently general.
9. Hence, an adequate methodological account of X must be *maximally* general in the sense that it expresses what all the various historical cases of acceptable reasoning concerning X have in common.
10. Hence, methodological progress occurs by discovering counterexamples to our methodological accounts and by proceeding to find more general accounts which can handle those counterexamples plus the domain which the defective account already covered successfully.
11. (Corollary) Careful study of the history of science is essential to methodology.

An H-D methodological strategy, in which historical cases are used as tests of methodological claims, is dangerous if every claim is open to falsification and rejection except, possibly, the assumption that there exists a general methodology which unifies all sciences. A general, lawlike claim about nature is falsifiable in principle by a single counter instance. Although methodological claims often have substantive content, it is rarely defensible to treat them as falsifiable, unqualifiedly universal, lawlike claims. Given that the content of methodological rules or procedures frequently is domain specific, we should ask not, 'Are these procedures true or false as universal generalizations?' but rather, 'Where do these problem-solving methods apply?' Perhaps a more generative, less H-D strategy for doing historical methodology is advisable. At any rate, bringing the great diversity of research methods under a single account represents a false economy and a false robustness. The questions which the methodologist should ask are not 'What single account of scientific research is best?' and 'What does all research have in common?' but rather, 'Given a certain type of problem to solve, which search strategies are most efficient, or at least efficient enough – or at the very least promising?' and 'Given a problem-solving strategy, to which kinds of problems can it be successfully applied?'[43]

I do not wish to impugn the use of history of science in methodology, only the just-described use of it. Clearly, the appeal to contemporary science is also fraught with the dangers reflected in the argument above, where 'maximal generality' is extended to include present and future cases of acceptable scientific reasoning – and even possible but unactualized cases of acceptable scientific reasoning.

Instead of using empirical information about historical and contemporary scientific research to determine which research strategies work best (or well enough) in which contexts, this line simply takes for granted that there is one, general methodology which correctly describes and/or prescribes everything. This assumption, together with our increasing knowledge of actual scientific cases, makes historical (and contemporary) methodology still more abstract than the intuition-based methodologies of old. Ironically, under the sway of the generality presupposition, empirical study of science results in methodologies which are more general and more vapid – that is, more empty of empirical content – than ever before!

## 5. . . . and back again

So far I have been railing against overly-general philosophy of science. But, ironically, in one respect such general philosophy of science has not been general enough! Because his work is so well known and has been so influential, I shall continue my running example of Popper as representative of the kind of approach taken by a significant number of philosophers. In a sense, Popper fails to address the central problem of methodology: What reasons are there for thinking that the methods espoused are the best ways, sufficiently good ways, *or even possible ways* of achieving one's stated goals? Similar criticisms can be made of classical inductivists, Bayesians, *et alia*.[44]

General though Popper's and dozens of other methodological accounts are, they too often neglect questions of *general methodological strategy*, the matching of goals to methods for attaining them and the matching of both to the particular problem at hand. Given that there are an infinite (or indefinitely large) number of interesting theories on any given topic, the simple falsificationist strategy of dreaming up one hypothesis after another and testing it until we determine its falsity is not a strategy likely to lead us to the true theory, Popper's stated goal.[45] Even though Popper's methodology is self-correcting, it is not self-correcting in the right sort of way to make goal attainment likely. At the most general level, there are only a few basic search strategies available – plus manifold combinations and variations of these. Philosophers devote too little attention to the comparative merits of the different strategies in relation to the goals of inquiry and the nature of the search space. Adopting an AI perspective on research helps to bring such questions into focus.

Notice that if Popper and other general methodologists addressed these general strategy questions, they would almost surely be led to the more content- and context-specific questions which I advocated at the close of the preceding section: which kinds of strategies work (adequately or best) for which kinds of problem contexts? Once we localize our methodological concerns in this way, in turns out that the very methodological ideas (eliminative induction, H-D, Bayesianism, etc.) which are doomed to fail as general methodologies, have interesting applications in specific problem contexts. Thus, eliminative induction, Hempel's theory of confirmation, Popper's methodology, and Bayesian inference – in combination with domain-specific and 'expert systems' constraints – all have been incorporated into recent AI programs.[46] Here the comforting thought occurs to us that some widely rejected philosophical tools and models might find useful service as components of domain-specific problem-solving strategies.

### 6. From anti-circularity to bootstrap

This and the final section address what I consider to be two profound developments in contemporary philosophy of science. Unfortunately, I have enough space only to sketch the problems that must be addressed and to refer to some promising work of others.

In the opening pages of his well-known book, *Perception,* H. H. Price assures us that physiology can cast no light on the philosophical problems of perception:

> Since the premises of Physiology are among the propositions into whose validity we are inquiring, it is hardly likely that its conclusions will assist us . . . In any case, Science only professes to tell us what are the *causes* of seeing and touching. But we want to know what seeing and touching themselves *are*.[47]

We may call this the 'circularity' argument for the conclusion that philosophers need not know science. As Dennett observes, at best Price's argument holds water only if we are attempting to build up a foundational system of knowledge from scratch, against absolute skepticism.

A similar circularity argument underlay philosophy of science until fairly recently: detailed knowledge of real science is not necessary, certainly not knowledge of *historical* science, for philosophy of science is a *normative* (alternative, a logical or purely analytic) discipline. To suppose that learning how science actually has been practiced can inform methodology is to beg the question. A related argument is that scientific methods could not themselves be scientifically justified, upon pain of circularity and relativism. Hence, methods must be empirically neutral; they are 'logical' rather than 'empirical'. Hence, methods cannot be domain specific in the empirically committed sense (although they could be in the sense that mathematical methods can be domain-specific). Hence, methodology is akin to metalogic.

Charges of vicious circularity are common against pragmatic philosophical programs which are today taken quite seriously (or should be, in my judgment). For example, Dewey's view, recently echoed by Shapere,[48] that nothing is given, that even logic and methodological principles are developed in the course of inquiry and legitimated by that inquiry has been attacked as question-begging on the ground that legitimation presupposes the existence of some of those very principles. But even if those principles are objective truths, independent of all inquiry, they still had to be discovered and legitimated by inquiry. There is no other option short of classical apriorism.

Reference to apriorism reminds us of well known objections to W. V. Quine's attempt to eliminate the *a priori*-empirical distinction, thus making logic itself, in a broad sense, an empirical enterprise (and also undercutting the Kantian epistemology mentioned above).[49] Does not the justification of logical laws already presuppose a logic? The justification of inductive logic and methodology has been embroiled in circularity disputes for centuries. A recent case is the circularity charge against Nicholas Rescher's attempt to justify theory procedurally in terms of method and method in terms of what works in practice.[50] For is not 'what works' at bottom an inductive justification, so that inductive methods are themselves being inductively 'justified?'

All of these examples concern pragmatic attempts to reject a basic dualism or dichotomy in the philosophical tradition in favor of a 'bootstrap' conception of inquiry and its guiding principles. In each case I believe that the pragmatists are right, but I cannot attempt to defend their views here.

Another circularity objection underlay the 'absolute' observational-theoretical dichotomy and the conviction that all propositions, regardless of context, can be neatly ordered according to their grades of theoreticity (or, inversely, according to their epistemic priority). Justification must always be of the more theoretical claim by the less, upon pain of circularity or even meaninglessness. The data function as 'unjustified justifiers'. Hence, the confirmation of a theory cannot depend on 'higher' theoretical assumptions, much less (God forbid!) employ tenets of that selfsame theory in the premises of the justificatory argument.

Today, most philosophers hold a quite different view, one which reflects scientific practice rather than classical philosophical ideals. It is widely recognized that the 'data' themselves, the etymology of 'datum' notwithstanding, are not given but are the product of complex processes of reasoning and judgment in which theoretical assumptions have a central role.[51] The data are not prior to all theoretical assumptions; rather, there is a *symbiosis* of theory and data. Theory has a major role in determining (1) what can be measured or otherwise observed as data in the first place, (2) which data are relevant, (3) how the raw data are processed, (4) how the data are interpreted, (5) how the interpreted data are legitimately used in construction, confirmation, etc. In the typical case, some of the theoretical background is well established but additional, more conjectural assumptions are necessary. The data may in turn support some of those very assumptions. The term 'symbiosis' is employed by Pickering, who provides particularly interesting examples of the interpenetration of theory and experiment. Symbiosis represents still another form of bootstrap. Harry Collins has, in effect, shown how the calibration of complex test instruments (e.g., 'tuning' an atomic accelerator) involves similar bootstrapping.[52] Since the data and the methodological rules, and not only theories and instruments, must be bootstrapped into existence, science is much more a bootstrap affair than any traditional philosophers, whether empiricist or rationalist, recognized.

As for theory testing, Glymour has given a technical account of scientific testing and confirmation which he actually calls the 'bootstrap theory'.[53] By this he means that one part of a theory may be assumed in the test of another part, or even of itself, provided that two independent derivations of the result are available, so that cross checking (i.e., a genuine test) is possible.

These days, everyone employs the bootstrap metaphor, although the idea was glimpsed as long ago as Bacon and by several thinkers since, as I indicated in Section 3g. Science somehow lifts itself by its bootstraps. Its results mutually reinforce one another, its methods mutually reinforce one another, its methods help justify its results, but those results in turn help license those methods.

How then does bootstrapping avoid vicious circularity?

## 7. From linear justification to consilience, robustness, and reliability

I wish to apply the bootstrap idea to theory of justification – here only in a very sketchy way. As a field of science matures, its results become progressively more self justifying, in a sense. Once a critical mass of reliable results is available, these results provide a (fallible) basis for cross-checking each other and for justifying further results. To the extent that they are interderivable or even mutually supporting, they provide a certain amount of self-support. (In a wide sense, of course, all support is self-support, since even the data claims are part of the picture and require support themselves.) A mature science possesses far more powerful justification methods than a nascent discipline, which must rely heavily on general cultural knowledge in the absence of many technical results. The very fact that its independently produced results hang together in the right sort of way provides a certain amount of self-support. William Whewell termed this 'consilience of inductions' and held it to be a particularly important form of justification.[54] Actually, I am going beyond Whewell to incorporate the consilience idea within a bootstrap conception of justification.

At a minimum, bootstrap conceptions reject the once standard view that scientific justification is linear and foundational. Here we need to distinguish epistemological linearity from linearity in a purely logical sense. Logical linearity arises out of the conception of logical proof as a sequence of sentences, each one of which follows by means of a valid rule of inference from prior lines – and prior lines only. In other words, a claim can be justified only by logically prior claims and can, in turn, justify only posterior claims. No two claims can be mutually supportive, even a little, upon pain of circularity. Hence, the prohibition of circular reasoning lies behind the demand for, and the attraction of, logical linearity. Applied to an entire branch of knowledge, logical linearity induces a tree structure: a given claim is derivable from (or perhaps inductively supported by) a certain number of premises; each of these premises in turn is supported by further premises; and so on, for as far as we can extend the tree. This is commonly called the 'Euclidean' model of rational justification. On this model, every well-reasoned discipline has a Euclidean structure.

Epistemological linearity is similar but involves the further claim that there is a natural, epistemic ordering of propositions – and the corresponding requirement that a statement can be justified only by prior statements and can justify only posterior statements in the ordering. Logical linearity by itself determines no epistemic starting point. In Aristotelian terms, the argument can begin with 'what is better known in nature' as easily as it can begin from 'what is better known to us'. Logical linearity applies equally to the *ordo essendi* and the *ordo cognoscendi*.

Epistemological linearity is not so unconstrained. It is concerned with justification and requires that good justificatory reasoning ultimately must proceed from what is best known to us to what is less well known to us. The empiricist version of this requirement is well known: Justification begins from the bedrock data, upon which dispositional statements rest, and so on through succeeding grades of theoreticity.

This conception of scientific justification was destroyed by the rejection of an absolute, all-purpose, epistemological, observational-theoretical distinction. For this distinction, in the form which recognized a quasi-continuum of grades of theoreticity, amounted to the assumption that there is a linear, epistemic ordering of scientific claims. In fact, there is no universal epistemic order to be faithful to. As Putnam said, 'justification in science proceeds in any direction that may be handy'.

'But' – it will be objected – 'although there is no universal, epistemic ordering of propositions, in particular research contexts such an order will exist and must be honored'. True, a rough ordering will be important in specific contexts (although some theoretical claims will be better founded than some experimental claims); but this order does not entail that all justification must be linear, that only stronger claims can be used to support weaker claims, and never vice versa. In practice, scientists frequently appeal to less well certified claims to provide modest further support to better established claims. Radical replacement of the old, linear view of justification is necessary.

'But even if *epistemic* priority provides no general justification strategy, I can still retreat to *logical* linearity', the objector responds. 'Does not the logical argument itself impose a linearity on any justification enterprise? We must reason from premises to conclusions, justify those premises by reasoning to them as conclusions of previously unused premises, and so on back as far as the demands for justification remain serious. The resulting justification tree is linear in the sense that no statement can appear at more than one place along a single path through the tree – upon pain of circularity. Even Glymour's bootstrap does not transgress this inviolable principle. Surely you are not trying to persuade us, after all these years, that circularity is a virtue?'

Well, yes and no. Of course I do not deny that common forms of circular reasoning remain vicious. But it seems to me that (a) if we are to become utterly *serious* in our fallibilism, (b) if we are *seriously* to challenge the traditional, linear, Euclidean model of justification, (c) if we are really sincere in saying that all justification ultimately involves a proper sort of coherence among our claims, (d) if we are genuinely committed to a naturalistic epistemology, (e) if we take to heart the discovery that domain-specific, empirically committed methods can be stronger than global methods, and if the requirement of logical 'linearity' is tantamount to forbidding any closure, any circularity, however broad, in our reasoning, *then* we must allow for the possibility of a *virtuous circularity*.

Surely it often happens that each of a set of statements supports the credibility or reliability of the others. Roughly, the difference between virtuous and vicious circularity is that vicious circularity is too simple and direct: it is 'short circularity'. for it 'short circuits' the inferential relations in such a way that reliability and robustness are not improved.

To permit virtuous circularity is not to give up all pretense of reason, it is not to allow anything at all to be trivially provable. As Friedman points out, the sort of virtuous circularity that we are talking about does not guarantee that our claims are coherent; hence, it is quite different from vicious circularity.[55] After all, philosophers have long treated *coherence* ideas with respect (including nonlinear coherence), and a sophisticated sort of coherence is what we have in science, as Quine and many other writers have emphasized. Nor would this be the first departure from rigid adherence to classical logical models of rationality and justification. Inconsistency, that worst of logical sins, is today widely recognized as a frequent characteristic of scientific theorizing (and of the set of beliefs held by virtually every individual and community). Many scientific theories have been plagued by inconsistencies and near-inconsistencies or 'conceptual blowups', yet several such theories have been extremely fruitful.[56] In some cases the inconsistencies were contradictions-in-terms which disappeared with conceptual shifts (e.g., Darwin's theory of the evolution of species, Freud's theory of the unconscious mind). But even where the inconsistencies persisted, scientists were able to confine and control them; scientists certainly did not seize the opportunity afforded by elementary symbolic logic to show that such theories entail absolutely every claim and hence are completely useless. On the contrary, they employed the theories in all the usual scientific ways and as guides to improved theories which might be free of such defects.

William Wimsatt, developing insights of R. P. Feynman, R. Levins, and others, in effect attempts to carry bootstrap thinking beyond Glymour into the 'nonlinear' domain.[57] Although I do not agree with everything that Wimsatt says, his discussion of robustness, redundancy, reliability, nonlinear (nonserial) organization, 'Babylonian' vs. 'Euclidean' logic and mathematics, and heuristics is wonderfully suggestive. His emphasis on reliability rather than truth strikes me as just right. This means seeing scientists as more like engineers and less like Euclidean mathematicians – an insight confirmed by observation of scientists at work and by taking into account their points of view. Very roughly speaking, confirmation theory becomes a branch of reliability theory.

I conclude with a vision which I think is worth precise articulation. That Whewell could see the importance of consilience reflects the fact that various sciences had, by mid-19th century, matured to the point that the multiple determination (not just discoverability but multiple discoverability) of results was now possible, even at higher theoretical levels, and the fact that this was becoming evident even to philosophers and historians. Baby disciplines have too small bodies of results and established practices to provide powerful justification for their results, let alone self-sufficient justification. They must rely heavily on knowledge and technique available in the wider culture. But as sciences mature, they become more and more self-sufficient, more and more 'self-supporting'.[58] There remains a residual dependence on cultural notions of reliable observation, what is reasonable, etc., but the more interesting justifications become technically esoteric.

Any attempt to make a baby discipline self-supporting fails utterly, for the small body of 'knowledge' available means that the effort will result either in vicious circularity or in the unfounded justification of some claims in terms of other claims which cannot in turn be justified scientifically. By contrast, a mature discipline (perhaps together with its neighboring disciplines, depending on how finely we draw the intellectual grid) includes a sufficient mass of well-established theoretical and experimental results plus a sufficient body of standard practices to permit extensive cross checking of results at all levels, including cross-*level* checks (e.g., of experimental results by theoretical expectations and of methods against their results). Taken together, the justificatory arguments are circular, but (1) the radius of the circle is now large enough that many results are interrelated by a given argument, not just two or three and (2) a typical claim is backed by multiple arguments from different sets of premises rather than by just one. In short, any given argument provides only partial support; and the arguments collectively, while circular, are various. There is not just one, big, circular argument. And finally, the resulting coherence is a wonderful achievement and not at all guaranteed in advance.

The existence of virtuous circularity explains how it is possible to be a thoroughgoing fallibilist (anti-foundationist) and yet 'rationally' espouse a body of 'positive scientific claims'. Briefly stated, accepting fallibilism entails that one must adopt a coherence theory of justification, which is ultimately nonlinear, hence, non-Euclidean. Thus, linear logical structures have only a local, not a global, application.

Euclidean logic is rather like Euclidean geometry in its limitations. Local regions are safely treated in a Euclidean manner, but globally, the space of reasons ultimately curves back on itself. In the real world we need a 'spherical' logic just as we need a spherical geometry.

I conclude that there is virtuous as well as vicious circularity! After all, how can a hand pull up a boot by the strap unless the boot does its part by pushing up the hand?

## Notes

1. My title is not intended as an allusion to Alan Musgrave's interesting paper, 'Method or Madness' (1976). I thank Jim Woodward for stimulating conversations on these topics. He is not responsible for my excesses. My work was supported by the U.S. National Science Foundation.
2. Compare Agassi (1963).
3. Reichenbach (1939), Chap. 1. See Nickles (1985b) for a fuller discussion of method and methodology.
4. Koertge (1979), p. 246.
5. Rescher (1976).
6. See Nickles (1984a and b, 1985a); Dorling (1973); Glymour (1980); and Worrall (1982).
7. See Nickles (1972, 1976).
8. See Gilbert and Mulkay (1980) and Pickering (1984a and b).
9. See for example Latour and Woolgar (1979) and Knorr-Cetina (1981).
10. Admittedly, the historical and the AI viewpoints differ radically from the working scientists' perspective in some ways. For instance, scientists are not historians, and the working scientists' viewpoint is notoriously unhistorical. (See the Pickering quotation in Section 4.) While I think a deep appreciation for the historicity of science is essential to philosophy of science, in this paper I emphasize the working scientists' point of view.

    As for AI, it is not that I think scientific methodology can be reduced to computer programs. Quite the contrary: I am firmly convinced that experienced judgment will always be crucial. Nevertheless, it is a healthy exercise to imagine programming a robot to be a scientist in a given field (better, a society of robots to be a scientific community). One reason is that this exercise immediately forces one to adopt the point of view of the working scientist. One then appreciates that problem-finding and problem-solving tasks are present at all stages of research and that discovery is 'ubiquitous', as artificial intelligencer Douglas Lenat (1978) nicely puts it. A problem-solving approach to the research process is unavoidable. One also must pay more attention to overall methodological strategy and economy of research than philosophers are wont to do. A further point is that methodological moves are frequently more mechanizable, more reducible to routine, than philosophers are willing to admit. Although rational judgment will never give way to an algorithm when momentous decisions are at stake, much of the detailed work within a given subspecialty becomes fairly routine. Thus, artificial intelligencers at Stanford University and elsewhere, in consultation with specialists in such areas as organic chemistry and internal medicine, have managed to reduce some of their difficult work to programmable routine, either heuristic or algorithmic (Buchanan, 1983). Artificial intelligencers have found it necessary to employ content- and context-specific methods in place of all-purpose, general problem-solving methods. Hence, their work will scarcely interest those philosophers who count as method only what all sciences have in common.

Since AI theorists are acutely aware of the magnitude of their problems and goals and of the limited means at their disposal, they cannot avoid an immediate concern with economy of research, heuristics, the mutual and efficient adjustment of means to ends, and with satisficing rather than maximizing. They are very sensitive to the nature of problems and problem (or solution) spaces. Finally, AI researchers attempt far more constructive 'methodologies' than the H-D, Bayesian, and other methodologies which monopolize philosophers' discussions. For my views about constructive methodology, see Nickles (1985a and b).

11. Dawkins (1976), p. 147.
12. Hutchins and Levin (1981).
13. See Nickles (1980e, 1981, 1982).
14. Kuhn (1962).
15. Feyerabend (1975); Popper (1959, 1962, 1972).
16. I do not deny the importance of confirmation theory. I am here calling attention to the neglected matter of heuristic justification. See Urbach (1978), Wimsatt (1980, 1981), Nickles (1980f).
17. At least not in the mature, physical sciences. See Collins (1975, 1981).
18. These matters are explained in Nickles (1984b, 1985a).
19. Simon (1957). See also Rescher (1976, 1984).
20. See Wartofsky (1980); Brown (1978); Nickles (1980c).
21. For example, Mahoney (1979); Nisbett and Ross (1980). Mahoney's results have been challenged.
22. Popper's H-D method of 'conjectures and refutations' is philosophically important chiefly because of his path-breaking fallibilism. As a description of either the research process or the justification of its products, the H-D method is so general and weak that it can furnish little insight into scientific inquiry beyond what was known by methodologists a century or three ago. Popper's treatment of the H-D method seems more a reaction to certain methodologies of the past (especially inductivism) than an attempt to provide a detailed account of the doings of real scientists, past or present. An 'external' motivation for Popper's account was to distinguish scientific inquiry from everything else, particularly pseudo-science and metaphysics.
23. Nagel (1961).
24. See Grene (1977).
25. Quoted by Duda and Shortliffe (1983).
26. Lakatos (1970); Worrall (1978); Zahar (1983); see my reply in Nickles (1985a and 1987).
27. Putnam (1962), p. 216; Putnam's emphasis.
28. Glymour (1980).
29. Bacon (1620); Peirce (1877), Section 1.
30. Dewey (1938), pp. 3–4; Dewey's emphasis.
31. Shapere (1980).
32. Peirce (1877).
33. Friedman (1979), pp. 370–71.
34. Butterfield (1931).
35. On homogeneous domains and their laws, see Nickles (1976b). Some writers minimize the demarcation problem by seeking methodologies of problem solving *in general*.
36. See Friedman (1983) and Kamlah (1984).
37. Nickles (1971, 1977).

64

38. Mynatt, Doherty, Tweney (1978).

39. Laudan (1981).

40. Pickering (1984b, p. 7; Pickering's emphasis). Compare American astrophysicist Samual Langley's attack, in the late 1880's, on the stereotypical, 'retrospective view' of the typical science textbook writer, who presents scientific progress as 'the march of an army toward some definite and', reported in Moyer (1983), p. 109. See Langley (1889).

41. Although I sympathize with Pickering's attack, it seems to me that there are contexts in which our views about reality can be employed to help explain the outcomes of experiments, without circularity, and, on a larger scale, why science has developed or 'progressed' in certain directions rather than others. Scientific explanation is certainly possible, and metascientific explanation is sometimes possible as well. This issue is largely responsible for the sharp division between realist philosophers and irrealist sociologists of science. Actually, circularity is sometimes permissible; see Section 6 below.

42. I here fudge the descriptive-normative boundary, as philosophers taking this line often do.

43. Philosophers never ask the second question. Yet, as historians and sociologists have documented, scientists often, in effect, have a solution in search of a problem – a set of problem-solving techniques which they wish to apply to a new set of research problems (Nickles, 1982). Compare Donald Campbell's Kuhnian question, 'In what possible worlds, in what hypothetical ontologies, would which knowledge-seeking processes work?' (1977, p. 12).

44. Here the impatient reader will object that I keep picking on H-D views while ignoring the more detailed and more interesting account offered by Bayesian probabilistic inference. I agree that Bayesian accounts are more interesting and more sophisticated than H-D accounts, if plagued by difficulties of their own. My point here, however, is that, for all their virtues, Bayesian accounts of science remain very general and provide little more insight into the details of the research process (as opposed to analyzing the confirmation relations of its products) than do H-D methods. This is not surprising, for, crudely stated, Bayesian methodology is a probabilistic version of H-D methodology (cf. Salmon, 1967, Chap. 7).

45. See Grünbaum (1978).

46. For example, the BACON programs of Simon et al. employ a form of inductive (data driven) reasoning (Langley, Bradshaw, and Simon, 1981); Kelly, Scheines, and Glymour (n.d.) employ Hempel's theory of confirmation; the PROSPECTOR program employs Bayesian inference (Campbell, Hollister, Duda, and Hart, 1982); Shapiro (1981) uses Popper's method of conjectures and refutations. Some of this work aspires to great generality.

47. Price (1932, p. 2; Price's emphasis; quoted by Dennett, 1978, p. 250).

48. Shapere (1980).

49. Quine (1951).

50. Rescher (1977).

51. Compare Shapere (1982); Pickering (1984a); Pinch (1984). A similar point can be made about practice.

52. See Pickering (1984a and b); Collins, 1982).

53. Glymour (1980).

54. Whewell (1847). Actually, there are several distinct senses of 'consilience' to be found in Whewell. See Laudan (1971).

55. Friedman's valuable paper (1979) was referred to my attention after this paper was written and submitted; fortunately, some last minute additions were possible, including a sharpened statement of this point. On the virtuous circularity of 'self-substantiating' science, see also Rescher (1978, p. 51).

56. See Nickles (1980a).

57. Wimsatt (1981). Friedman (1979) also says some important things about reliability.

58. Compare Shapere (1980) on the progressive internalization of science.

# Bibliography

Agassi, J. (1963), *Toward an Historiography of Science* (The Hague: Mouton).

Bacon, F. (1620), *Novum Organum*, trans. J. Spedding, R. Ellis, and D. Heath, 1863; reprinted in *The New Organon and Related Writings*, F. Anderson, ed. (Indianapolis: Bobbs-Merrill, 1960).

Brown, H. (1978), 'On Being Rational', *American Philosophical Quarterly* 15: 241–248.

Buchanan, B. (1983), 'Mechanizing the Search for Explanatory Hypotheses', in *PSA 1982*, Vol. 2, P. Asquith and T. Nickles, eds. (East Lansing, MI: Philosophy of Science Association), pp. 129–146.

Butterfield, H. (1931), *The Whig Interpretation of History* (reprinted New York: Norton, 1965).

Campbell, A., V. Hollister, R. Duda, and P. Hart (1982), 'Recognition of a Hidden Mineral Deposit by an Artificial Intelligence Program', *Science* 217: 927–929.

Campbell, D.T. (1977), 'Descriptive Epistemology: Psychological, Sociological, and Evolutionary', preliminary draft of the William James Lectures, Harvard University.

Collins, H. (1975), 'The Seven Sexes: A Study of the Sociology of a Phenomenon or the Replication of Experiments in Physics', *Sociology 9:* 205–224.

Collins, H. ed. (1981), *Social Studies of Science* 11 (February), special issue on the relativist program and on replication of experimental results.

Collins, H. (1982), 'The Function of Calibration in Experimental Science', read at the annual meeting of the Society for Social Studies of Science, Philadelphia.

Dawkins, R. (1976), *The Selfish Gene* (Oxford: Oxford University Press).

Dennett, D.C. (1978), 'Current Issues in the Philosophy of Mind', *American Philosophical Quarterly* 15: 249–261.

Dewey, J. (1938), *Logic, The Theory of Inquiry* (New York: Henry Holt).

Dorling, J. (1973), 'Demonstrative Induction: Its Significant Role in the History of Physics', *Philosophy of Science* 40: 360–372.

Duda, R. and E. Shortliffe (1983), 'Expert Systems Research', *Science* 220: 261–268.

Feyerabend, P.K. (1975), *Against Method* (London: New Left Books).

Friedman, M. (1979), 'Truth and Confirmation', *Journal of Philosophy* 76: 361–382.

Friedman, M. (1983), *Foundations of Space-Time Theories* (Princeton: Princeton University Press).

Gilbert, N. and M. Mulkay (1980), 'Contexts of Scientific Discourse: Social Accounting in Experimental Papers', in *The Social Process of Scientific Investigation*, K. Knorr, R. Krohn, and R. Whitley, eds., Sociology of the Sciences Yearbook 1980 (Dordrecht: Reidel), pp. 269–294.

Glymour, C. (1980), *Theory and Evidence* (Princeton: Princeton University Press).

Grene, M. (1977), 'Philosophy of Medicine: Prolegomena to a Philosophy of Science', in *PSA 1976*, Vol. 2, F. Suppe and P. Asquith, eds. (East Lansing, MI: Philosophy of Science Association), pp. 77–93.

Grünbaum, A. (1978), 'Popper vs Inductivism', in Radnitzky and Andersson (1978), pp. 117—142.

Hutchins, E. and J. Levin (1981), 'Point of View in Problem Solving', *Proceedings of the Third Annual Conference of the Cognitive Science Society*, Berkeley, pp. 200–202.

Kamlah, A. (1984), 'The Neo-Kantian Origin of Hans Reichenbach's Principle of Induction', read at University of Pittsburgh.

Kelly, K., R. Scheines, and C. Glymour (n.d.): 'HEMPEL: A Program for Machine Testing of Hypotheses of Any Logical Form', draft, University of Pittsburgh.

Knorr-Cetina, K. (1981), *The Manufacture of Knowledge* (Oxford: Pergamon Press).

Koertge, N. (1979), 'The Problem of Appraising Scientific Theories', in *Current Research in Philosophy of Science*, P. Asquith and H. Kyberg, eds. (East Lansing, MI: Philosophy of Science Association), pp. 228–251.

Kuhn, T.S. (1962), *The Structure of Scientific Revolutions* (Chicago: University of Chicago Press; 2nd ed., enlarged, 1970).

Lakatos, I. (1970), 'Falsification and the Methodology of Scientific Research Programmes', in *Criticism and the Growth of Knowledge*, I. Lakatos and A. Musgrave, eds. (Cambridge: Cambridge University Press), pp. 91–195.

66

Langley, P., G. Bradshaw, and H. A. Simon (1981), 'Bacon.5: The Discovery of Conservation Laws', *Proceedings of the Seventh International Joint Conference on Artificial Intelligence,* I., pp. 121–126.

Langley, S.P. (1889), 'The History of a Doctrine', *American Journal of Science,* 3rd Ser., 37: 1–23.

Latour, B. and S. Woolgar (1979), *Laboratory Life* (London: Sage Publications).

Laudan, L. (1971), 'William Whewell on the Consilience of Inductions', *The Monist* 55, 368–91; reprinted in Laudan's *Science and Hypothesis* (Dordrecht: Reidel, 1981), pp. 163–180.

Laudan, L. (1980), 'Why Was the Logic of Discovery Abandoned?', in Nickles (1980c), 173–183.

Laudan, L. (1981), 'A Confutation of Convergent Realism', *Philosophy of Science* 48: 19–49.

Lenat, D. (1978), 'The Ubiquity of Discovery', *Artificial Intelligence* 9: 257–285.

Mahoney, M.J. (1979), 'Psychology of the Scientist: An Evaluative Review', *Social Studies of Science* 9: 349–75.

Moyer, A. (1983), *Amerian Physics in Transition* (Los Angeles: Tomash).

Musgrave, A. (1976), 'Method or Madness?', in *Essays in Memory of Imre Lakatos,* R. S. Cohen *et al.,* eds. (Dordrecht: Reidel), pp. 457–491.

Mynatt, C., M. Doherty, and R. Tweney (1978), 'Consequences of Confirmation and Disconfirmation in a Simulated Research Environment', *Quarterly Journal of Experimental Psychology* 30: 395–406; excerpts reprinted in their *On Scientific Thinking* (New York: Columbia University Press, 1981), pp. 145–157.

Nagel, E. (1961), *The Structure of Science* (New York: Harcourt, Brace).

Nickles, T. (1973), 'Two Concepts of Intertheoretic Reduction', *Journal of Philosophy* 70: 181–201.

Nickles, T. (1974), 'Heuristics and Justification in Scientific Research: Comments on Shapere', in Suppe (1974), pp. 571–589.

Nickles, T. (1976a), 'Theory Generalization, Problem Reduction, and the Unity of Science', in *PSA 1974,* R. S. Cohen *et al.,* eds. (Dordrecht: Reidel), pp. 33–75.

Nickles, T. (1976b), 'On Some Autonomy Arguments in Social Science', in *PSA 1976,* Vol. 1. F. Suppe and P. Asquith, eds. (East Lansing, MI: Philosophy of Science Association), pp. 12–24.

Nickles, T. (1977), 'On the Independence of Singular Causal Explanation in Social Science: Archaeology', *Philosophy of the Social Sciences* 7: 163–187.

Nickles, T. (1980a), 'Can Scientific Constraints Be Violated Rationally?', in Nickles (1980b), pp. 285–315.

Nickles, T., ed. (1980b), *Scientific Discovery, Logic, and Rationality* (Dordrecht, Reidel).

Nickles, T. (1980c), 'Rationality and Social Context', in Nickles (1980d), pp. xiii–xxv.

Nickles, T., ed. (1980d), *Scientific Discovery: Case Studies* (Dordrecht: Reidel).

Nickles, T. (1980e), 'Scientific Problems: Three Empiricist Models', in *PSA 1980,* Vol. I, R. Giere and P. Asquith, eds. (East Lansing, MI: Philosophy of Science Association), pp. 3–19.

Nickles, T. (1980f), 'Heuristic Appraisal, Discovery, and Justification: Whither Philosophy of Science?', read at University of Pittsburgh.

Nickles, T. (1981), 'What Is a Problem that We May Solve It?', *Synthese* 47: 85–118.

Nickles, T. (1982), 'Notes on Problem Generation', unpublished draft.

Nickles, T. (1984a), 'Scoperta e Mutamento Scientifico' ('Discovery and Scientific Change'), *Material Filosofici* 10: 7–27.

Nickles, T. (1984b), 'Positive Science and Discoverability', in *PSA 1984,* Vol. 1, P. Asquith and P. Kitcher, eds. (East Lansing, MI: Philosophy of Science Association), pp. 13–27.

Nickles, T. (1985a), 'Beyond Divorce: Current Status of the Discovery Debate', *Philosophy of Science* 52: 177–206.

Nickles, T. (1985b), 'Methodologia, Euristice, e Razionalità' ('Methodology, Heuristics, and Rationality'), in *I Modi del Prograsso,* M. Pera and J. Pitt, eds. (Milan: Il Saggiatore), pp. 87–116.

Nickles, T. (1987), 'Lakatosian Heuristics and Epistemic Support', *British Journal for the Philosophy of Science,* in press.

Nickles, T. (in preparation), *Scientific Discovery: Classical and Contemporary Programs.*

Nisbett, R. and L. Ross (1980), *Human Inference: Strategies and Shortcomings of Social Judgment* (Englewood Cliffs, NJ: Prentice-Hall).

Peirce, C.S. (1877), 'The Fixation of Belief', reprinted in *Collected Papers*, Vol. 5, C. Hartshorne and P. Weiss, eds. (Cambridge: Harvard University Press, 1931–1935), pp. 358–387.

Pickering, A. (1984a), 'Against Putting the Phenomena First: The Discovery of the Weak Neutral Current', *Studies in History and Philosophy of Science* 15: 85–117.

Pickering, A. (1984b), *Constructing Quarks: A Sociological History of Particle Physics* (Chicago: University of Chicago Press).

Pinch, T. (1984), 'Towards an Analysis of Scientific Observation: The Externality and Evidential Significance of Observation Reports in Physics', *Social Studies of Science* 15, 3–35.

Popper, K.R. (1959), *The Logic of Scientific Discovery* (London: Hutchinson). Expanded translation of *Logik der Forschung* (Vienna, 1934).

Popper, K.R. (1962), *Conjectures and Refutations* (New York: Basic Books).

Popper, K.R. (1972), *Objective Knowledge* (Oxford: Oxford University Press).

Price, H.H. (1932), *Perception* (London).

Putnam, H. (1962), 'What Theories Are Not', in *Logic, Methodology and Philosophy of Science*, E. Nagel, P. Suppes, and A. Tarski, eds. (Palo Alto: Stanford University Press). Reprinted in Putnam's *Mathematics, Matter and Method, Philosophical Papers*, Vol. I, 2nd edition (Cambridge: Cambridge University Press, 1979), pp. 215–227.

Quine, W.V. (1951), 'Two Dogmas of Empiricism', *Philosophical Review* 60: 20–43. Reprinted with changes in Quine's *From a Logical Point of View* (Cambridge: Harvard University Press, 1953), pp. 20–46.

Radnitzky, G. and G. Andersson, eds. (1978), *Progress and Rationality in Science* (Dordrecht: Reidel).

Reichenbach, H. (1938), *Experience and Prediction* (Chicago: University of Chicago Press).

Rescher, N. (1976), 'Peirce and the Economy of Research', *Philosophy of Science* 43: 71–98. Largely reprinted, with revisions, as Chapter 4 of Rescher (1978).

Rescher, N. (1977), *Methodological Pragmatism* (Oxford: Blackwell).

Rescher, N. (1978), Peirce's Philosophy of Science (Notre Dame, IN: University of Notre Dame Press, 1978).

Rescher, N. (1984), *The Limits of Science* (Berkeley: University of California Press).

Salmon, W. (1967), *The Foundations of Scientific Inference* (Pittsburgh: University of Pittsburgh Press).

Shapere, D. (1980), 'The Character of Scientific Change', in Nickles (1980b), pp. 61–116. Reprinted in Shapere's *Reason and the Search for Knowledge* (Dordrecht: Reidel, 1984), pp. 205–260.

Shapere, D. (1982), 'The Concept of Observation in Science and Philosophy', *Philosophy of Science* 49: 485–525.

Shapiro, E. (1981), 'An Algorithm that Infers Theories from Facts', *Proceedings of the 7th International Joint Conference on Artificial Intelligence*, 446–451.

Simon, H.A. (1957), *Models of Man* (New York: John Wiley).

Urbach P. (1978), 'The Objective Promise of a Research Programme', in Radnitzky and Andersson (1978), pp. 99–113.

Wartofsky, M. (1980), 'Scientific Judgment: Creativity and Discovery in Scientific Thought', in Nickles (1980d), pp. 1–20.

Whewell, W. (1847), *The Philosophy of the Inductive Sciences*, 2nd ed., 2 vols.(London).

Wimsatt, W. (1980), 'Reductionistic Research Strategies and their Biases in the Units of Selection Controversy', in Nickles (1980d), pp. 213–259.

Wimsatt, W. (1981), 'Robustness, Reliability and Multiple-Determination in Science', in *Knowing and Validating in the Social Sciences: A Tribute to Donald T. Campbell*, M. Brewer and B. Collins, eds. (San Francisco: Jossey-Bass), pp. 124–163.

Worrall, J. (1978), 'The Ways in which the Methodology of Scientific Research Programmes Improves on Popper's Methodology', in Radnitzky and Andersson (1978), pp. 45–70.

Worrall, J. (1982), 'Mr Newton and Hypotheses', read at University of Pittsburgh.

Zahar, E. (1983), 'Logic of Discovery or Psychology of Invention?', *British Journal for the Philosophy of Science* 34: 243–261.

# Historical Realism and Contextual Objectivity:
# A Developing Perspective in the Philosophy of Science[1]

MARJORIE GRENE
*University of California, Davis*

The twentieth century ought to have been a century of philosophical reform – goodness knows, enough people have tried to make it so. But philosophical revolutions appear painfully difficult to accomplish. In the case of philosophical reflection about scientific knowledge, perhaps, one ought not to complain, since the orthodoxy we need to overcome was itself established *as* orthodoxy only in mid-century. Still, the situation is puzzling. Although so far as I know the so-called 'received view' is by now received only in a few outlying places, and although every reprint or preprint I receive seems to announce a new almost orthodoxy from yet another angle – in spite of all that, a positive, non-apologizing acceptance of a new approach to the philosophy of science seems to be still outstanding. When I have tried on a number of occasions in the past few years to suggest what such a new approach might look like, I have been received with the announcement that I was presenting a position of pure relativism, or, worse yet, pragmatism.[2] That there is a quasi-relativistic component in the new philosophy of science I am willing to admit, though I would prefer to call it a perspectival component (I will return to that briefly later). But pragmatism! If that is the kind of abandonment of any interest in cognitive claims we are supposed to be advocating, we are presenting our case very badly – or rather I have been doing so, for of course I am speaking only from my own experience. At first when I thought about what I wanted to say on this occasion, I considered making a stab at trying to reply once more to these misunderstandings, not only of my own view, which does not matter much, but of the general view I see developing. But then the materials tending toward a new positive position seemed to be coming in so fast, that I wanted chiefly to report them, to try to offer a sense that something promising really is happening, at long last, post-positivism and, just as emphatically, post-incommensurability, in the development of a new approach to the philosophy of science. In fact I suppose what I shall be attempting is a kind of compromise: to report what I have noticed lately (by no means a review of the literature, but an anecdotal account of some bits that happen to have come my way), and to do so with some comment in the shape of something like an argument, to give a bit of philosophical ballast to my otherwise chatty sketch.

But where to begin? If I started chronologically, I would have to take off, not from

*Nancy J. Nersessian (editor), The Process of Science. ISBN 90-247-3425-8.*
© *1987, Martinus Nijhoff Publishers (Kluwer Academic Publishers), Dordrecht. Printed in the Netherlands.*

a reprint recently received from Nancy Nersessian[3], nor from Lee Rowen's dissertation, of which her supervisor wrote me[4], nor from Philip Kitcher's *Philosophical Review* paper of 1984[5] or his recent 'Darwin's Achievement'[6], nor from John Beatty's study of Darwin on species, 'What's in a Word?'[7], nor yet from Richard Burian's paper in the Depew and Weber volume recently published (or some other papers in that volume)[8], nor from Sober's new book[9], nor, indeed, from any of the current literature, but from Michael Polanyi's *Personal Knowledge* of 1958.[10] So far as I know, that was the first well-articulated statement of the position that is just now being so painfully reconstructed by so many people quite independently both of it and in large part of one another. But as I have heard a fellow philosopher remark, its rhetoric was wrong. People didn't in those days – pre-Rorty or Nozick? – take to high-toned sermonizing philosophy. When I asked Polanyi – who was then about the age I am now – why he was so much more advanced than all those clever young men in Oxford, he answered promptly: 'Of course, because I've been advancing longer!' Now that I have reached the season of aged precocity, however, I am glad to notice that the young – a number of them – are advancing, too. But before I remark on some of these, I must correct my chronology again. Basically, though not with a special view to the philosophy of science, Polanyi's position had been adumbrated, in an almost equally unfashionable philosophical style, in Merleau-Ponty's *Phenomenology of Perception,* way back in 1945.[11] And Merleau-Ponty's philosophical perspective, though again in a very different language, and with a very different center of attention, is once more paralleled, or confirmed, with a new discipline in experimental psychology to back it up, in J.J. Gibson's last book, *The Ecological Approach to Visual Perception,* published a few months before his death in 1979.[12] So 1958, 1945 and 1979 could provide us with a fuller *biblion* to furnish principles for the new developments.

Not that they *will* do so. The Gibson work, possibly; there *are* a few philosophers thinking ingeniously about that, though not in the direction of philosophy of science. In general, however, it takes – as it does in all disciplines – a current *community* of thinkers to permit the effective formulation of a new perspective; isolated anticipators make very little difference. And in this case such a community is still in the making. Thus, in general, each of the handful of recent writers I have mentioned and will mention again comes on as relatively isolated, considering him(her)self rather as fighting an Establishment than as establishing, let alone subscribing to, one.

Lee Rowen does acknowledge some sources, one in particular from epistemology. Let me start with her. She was trained in biochemistry at Stanford, and turned to philosophy with the kind of experience of ongoing work in science that is precisely one of the striking features of the new perspective. Instead of logical reconstruction almost wholly detached from any contact with the life of science, this is reflection about a specific family of practices, the critical examination of a tradition from within that tradition itself. The question is, how, within a given field of inquiry, a group of critical truth seekers move forward to state justified beliefs, which they hope are true, in conformity with standards of what Rowen calls *epistemic responsibility*. This seems to me a seminal concept. Someone (I have forgotten who) said it was already current; I don't know where, but if someone can tell me, I shall be happy to acknowledge its source.[13] The question is: how, with a due sense of what the important problems are and a just weighing of evidence, do scientists move to the solution of problems on the forefront of their respective subdisciplines? Thus the analysis of epistemic claims is placed where it belongs, within a human and *therefore* rational context. So instead of being demoted from the intense inane of pure logical reconstruction to the literal non-sense of sociology of knowledge in its so-called 'strong' form, the claims of scientists to seek and to discover truths about the natural world can be critically assessed as the reasonable but corrigible claims they are.

Rowen's case-study is of the chemical coupling theory of oxidative phosphorylation. I quote:

In 1953 when the first 'biochemical theory' (the chemical coupling mechanism) was proposed by E.C. Slater, it would be fair to say that 'ox-phos' was viewed as a difficult but typical biochemical research problem, and hence solvable by standard biochemical methodological practices. This view turned out to be essentially wrong. To see that it was wrong, and that a competing 'radical fringe' theory (the chemiosmotic mechanism) based on a novel postulation of the appropriate mechanistic model for 'ox-phos' was right required the development of new methods and concepts.[14]

I shall not try to report further on this project here, but recommend that my readers look for Rowen's work whenever it is eventually published. I want here instead to consider the epistemological source she relies on, which, allergic to most recent epistemological literature as I confess to being, I would never have noticed but for her acknowledgement.

That source is a paper published by David Annis in the *American Philosophical Quarterly* on a position he calls 'contextualism'.[15] He is concerned with the controversy about epistemic justification, insisting that the justification of a statement can be properly called for only by serious objections in a fitting context. Thus if Jones drops by a party and looks to see if his friend Smith is there, glancing about in the group, and asking another guest or two, will give him adequate evidence. But if he is a detective looking for Smith the alleged murderer, such a hasty inquiry would not suffice to justify his belief. Granted, Smith is either there or not there. There is a sense in which the truth of the matter transcends justification. That is why I referred earlier to 'critical truth seekers . . . stating justified beliefs, which they *hope* are true'. There is plenty of acceptable knowledge about, both in learned or scientific disciplines and in ordinary life; we can assert or, as we constantly do, implicitly rely on, such knowledge without any foundation-based, 150% guarantee of its infallibility. In other words, we can reasonably claim to know a great deal without claiming any ultimate and indefeasible ground for knowing that we know. We hope we know we know, just as we hope we know. Or perhaps one could say that there is a benign regress from knowing to knowing that we know, and so on and on, a regress that can take at any time a contextual cutoff, as it has a contextual source. But the point is that the regress is benign just because at any level inquiry is taking place *within a real world*. To understand this we need to note two moves in Annis's – and Rowen's – argument. They correspond to the two tags of my title: *historical realism* and *contextual objectivity*.

*First, then, historical realism.* In his paper, Annis talks about 'naturalizing' justification, but he means 'socializing'. And since in my view the notion of a *naturalized* epistemology has entailed chiefly an abandonment of epistemology altogether in favor of some empirical discipline, whether physics or psychology, I prefer not to use that suspect term. One doesn't need it; what Annis means is clear enough. In fact, he applies his thesis specifically to philosophy of science, and even talks as though the reform I find to be painfully under weigh had been already accomplished a decade ago. He writes:

> Positivists stress the *logic* of science – the structure of theories, confirmation, explanation – in abstraction from science as actually carried on. But much of the main thrust of recent philosophy of science is that such an approach is inadequate. Science as *practiced* yields justified beliefs about the world. Thus the study of the actual practices, which have changed through time, cannot be neglected. The present tenor in the philosophy of science is thus toward a historical or methodological realism[16]

That is my source for the phrase 'historical realism'; 'methodological' seems to me a bit too skimpy. And I would want to say, not just 'science as practiced yields . . .', but 'science as practiced *has yielded* justified beliefs about the world'. In any ongoing scientific discipline there is a body of truths (as we hope) already relied on, within the compass of which inquiry at the forefront of knowledge proceeds. What used to be problematic becomes accepted knowledge; what *was* theoretical comes to belong to the background of accepted 'fact'. Indeed, one of the intriguing questions in the new philosophy of science is the problem of how theories become facts – an inconceivable transition in the older tradition, but a fundamental one on this newer, more concretely realistic (in *that* sense if you like more 'naturalistic') style of thought.[17] But in any case, I cannot try to go into that question here, even supposing that I knew (as I don't) how to set about it. I leave it open for somebody else to think about. Here I just want to stress the *realism* in historical realism: the practices of critical truth seekers are truly the practices of *truth* seekers. Scientists are people really puzzled about how something in the real world really works – say, about the cause of shell removal in black-headed gulls or the structure of a particular hexoprotein. Granted, their puzzles arise within a framework of beliefs and techniques that is highly acculturated, but after all labs – let alone gull's nests – are real places, too, using real equipment and real experimental material to ask real questions. As J.J. Gibson put it:

> The human environment is not a *new* environment – an artificial environment distinct from the natural – but the same old environment modified by man . . . It is a mistake to separate the natural from the artificial as if there were two environments: artifacts have to be manufactured from natural substances. It is also a mistake to separate the cultural environment from the natural environment, as if there were a world of mental products distinct from the world of material products. There is only one world, however diverse, and all animals live in it, although we human animals have altered it to suit ourselves. We have done so wastefully, thoughtlessly, and, if we do not mend our ways, fatally.[18]

In this context (of the unity of the natural world) I may even claim here a place for myself as one of those rightly neglected anticipators. In my research lecture at Davis more than a dozen years ago, really reporting on Helmuth Plessner's philosophical anthropology (yet another anticipation!), I ventured to present both a (sort of) definition and a diagram of the human condition (see Figure 1).[19]

A (human) individual, I said, is a personification of nature through participation in, and as expression of, a culture. In other words, a human being is a natural entity which achieves personhood through immersion in a culture; but a culture is in turn a modification of nature *within* nature. There is no wholly extra-natural stance we human beings can take. Even the highest high tech uses *some* natural materials and depends ultimately on a natural environment for its existence. Just apply my general model, then, to the individual scientist in his(her) lab and you will see, I think, why historical realism in philosophy of science is utterly, almost *a priori,* realistic in starting point and outcome.

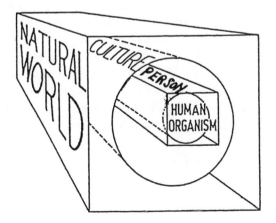

*Fig. 1.*

In short, we really must get over the absurdity of holding what is cultural to be *unreal*. After all, acid rain is real, smog is real, the bomb is real. And if ostrich-like we choose to ignore the evils that are our own doing, our ignorance is real.

But to get back to Annis, historical realism and the philosophy of science: I must make two further digressions. As Lee Rowen acknowledges Annis as her source for 'contextualism', so Annis in turn, in the context of philosophy of science, acknowledges Fred Suppe's long introduction and ' '76 Afterword' to his *Structure of Scientific Theories*[20] as the source of his (Annis's) optimistic statement about 'the present tenor in the philosophy of science' – as of 1978! I have read Suppe's original introduction to that volume, but found it too painfully and laboriously concerned with the old orthodoxy; still, perhaps we ought to count it (or at least the 'After-word') as yet another anticipation of what was, and largely still is, to come.

On the other hand, writing more recently, another of those *I* find contributing to the new philosophy of science, Richard Burian, in his paper 'On Conceptual Change in Biology' starts out, to my surprise, by declaring philosophy of science to be just now – in 1985! – at a most distressing impasse, and in particular to be plagued by an unhappy breakdown of the once ready solution to the 'problem of scientific realism' – that is, the problem of the 'reality' or otherwise of theoretical entities.[21] Non-sense! There is no global problem of 'scientific realism'. It was a pseudoproblem, if ever there was one, arising within the context of the 'received view' or its ancestor, positivism. In the first decade of the century, perhaps, there was still a context within which it made sense to ask whether atoms were real, as Emil Fischer, a 1902 Nobel Laureate, in fact did in his Faraday lecture. And in those days perhaps it made sense still to conceive of scientists, as Mach had done of himself and others, as spending their professional lifetimes counting their sensations. Yet *we* can now see how these were errors – both in the case of the atom and of the positivistic claim in general. In *our* intellectual context, we can be, straight off, with historical hindsight, knock-down-drag-out realists. We don't *have* a 'problem of scientific realism' – and neither, by now, should Burian. Once we assimilate the position articulated by Annis and exemplified in Rowen's account, we can start out by asserting confidently, as I have already done, that scientists are real live people concerned with real puzzles about how, and why, certain processes or events happen as they do in the real world. Only the legacy of logical positivism permitted such a so-called problem as that of 'scientific realism' to be formulated, let alone resolved. Instead we can now start with a global, or, as I've sometimes called it, comprehensive realism. Here I have adopted Annis's phrase 'historical realism' in order to stress the hoped-for contact with reality inherent in the historical practices of certain, in this case scientific, communities.

At the same time, as Annis goes on to insist in the passage I have already quoted:

From the fact that justification is relative to the social practices and norms of a group, it does not follow that they cannot be criticized nor that justification is somehow subjective. The practices and norms are epistemic and hence have as their goals truth and the avoidance of error.[22]

Thus within the practice of a group, norms are followed in the ascertainment of 'fact', in the weighing of evidence, in the formulation and testing of causal explanations. We have what I am calling *contextual objectivity,* the only kind of objectivity there is. But it's not too bad. It is neither arbitrary, nor irredeemably authoritarian, but entails the possibility of criticism. Again, such criticism must be 'serious', 'the result', as Annis puts it, 'of some jar or hitch in our experience of the world'. But, he continues,

> such objections will always be local rather than global. Some practice or norm and our experiences of the world yield the result that another practice is problematic. A real objection presupposes some other accepted practice.[23]

Here I must remark that that is just the position of *Personal Knowledge.* Insofar as people read that book, they noticed chiefly the emphasis on incommensurability between cognitive communities, for instance in the case of the Azande as against their European observers, but not the more basic bipartite thesis: (1) that we can and, in our scientific and scholarly traditions, *must* learn to be critical of our own fundamental (methodological and ontological) beliefs, even though (2) we can criticize such beliefs, as Annis puts it, 'only locally'. Since we *are* as human persons expressions of a culture, we can no more abandon totally *all* our norms of knowing and doing than we can totally abandon ourselves. But that does not mean that we need abandon altogether the cultivation of our critical powers.

The difficulty of understanding and accepting this conception stems partly, I suppose, from our lingering commitment to the split between subjectivity and objectivity characteristic of the modern tradition. But there is no such split. Babies don't start as pure egos, and add a hypothetical 'external world': they start very early attending to objects and events which they learn to discriminate. Such learning is needed for achievement of the detailed, contextual objectivities characteristics of adult, including scientific, life; but at the same time there is a kind of vague global objectivity: a being-over-against-out-in-the-world to begin with, and it is within that global thereness-as-otherness that mentality develops. As the passage I have quoted from J.J. Gibson indicates, this is an *ecological* realism, within which history, and our kind of cultural context, appear at the human level.

Contextual objectivity, then, is critical truth-seeking as practiced within a given social context. This conception, along with the historical realism with which it is associated, is illustrated by the other current contributions I mentioned earlier. None of them is as explicit, or comprehensive, philosophically, as Rowen-relying-on-Annis; but they are all moving, less or more explicitly, in the same direction. Nersessian's 'Aether-Or' in a recent issue of *Studies in the History and Philosophy of*

*Science,*[24] works from a case in the history of physics. My other instances, however, all come from biological case studies – and again, I make no claim to be citing all, or even a large fraction, of the available literature even in this limited field. But let me take off again for now from Burian's paper about conceptual change in the case of the gene: he in turn acknowledges his debt to Philip Kitcher's essay, 'Genes'[25] and to Kitcher's concept of 'reference potential',[26] but Burian goes further than Kitcher in his explicit analysis of this case and he differs from Kitcher on at least one important philosophical issue which I cannot go into here.[27] In this paper he is sketching some of the ambiguities and changes in usage of the term 'gene' in what may perhaps be roughly called the Bateson-Morgan-Benzer period. In general, he finds 'at least four ways in which the reference of a particular use of the term 'gene' or one of its cognates, might be specified'. 'Which one of these is relevant', he continues, 'will turn on the dominant intention of the scientist and the context of the discussion'.[28] These four ways are: 1st, conformity to conventional usage; 2nd, accuracy – this sometimes conflicts with conformity; and may also support distinctions forbidden by the third criterion, clarity, which demands Fregean fixity of reference; and fourth, there is what Kitcher has called 'naturalism', i.e. (in Burian's words) 'the intention to refer to the natural kind occurring or operating in a certain situation or in a certain class of cases'.[29]

Any or all of these criteria may be applicable in a given case. In particular, Burian concludes, he suspects 'that one must have recourse to naturalism over and above conformity, accuracy and clarity in order to put forth a successful account of the grounds on which Mendel, Bateson, Morgan, Benzer and all the rest may be construed as employing concepts of the same thing – the gene'.[30]

Burian is dealing with the development of a biological *concept* – and as Ernst Mayr has argued in his *Growth of Biological Thought*[31], that is an approach especially fruitful for the study of the biological sciences. More globally, a historically enriched study of developing scientific techniques and the methodological and ontological beliefs associated with them is found in Kitcher's '1953 and All That: A Tale of Two Sciences', (a study of molecular biology in relation to transmission genetics) as well as in his recent 'Darwin's Achievement'.[32] Of the meta-biological writers, indeed, Kitcher, along with Rowen, shows, in my view, the most acute sense of the very broad philosophical reform that is being accomplished through these concrete case studies. Both of Kitcher's papers are too rich, and especially the '1953' piece too technical, for me to try to summarize here. Again, I can only say: every student of the philosophy of science should read them.

However, there is one puzzle raised, for me at least, by Kitcher's '1953' paper, alongside Burian's and also along with Sober's book on *The Nature of Selection,*[33] and that concerns the place, in the new philosophy of science, of the problem of reduction. It is the old question of the possible reduction of transmission genetics to molecular biology with which Kitcher is explicitly concerned. (That was also the question around which David Hull had organized his textbook some years ago.[34])

Hence, in the Kitcher case, 'the tale of two sciences': once more he is asking, can the older, relatively macro- one be reduced to the new molecular discipline? The answer is no, but for very subtle reasons, which should give both reductivists and anti-reductivists some solace. Moreover, to return to Burian, one of the 'problems' he finds the 'received view' had solved (along with 'scientific realism') was that of theory reduction – and now it's come unstuck (in the concluding part of his paper he shows how); or again to take yet another example, the 'reductivism' theme runs, not like a red thread, but like a subtly blending rose-grey one, perhaps, through the elegantly articulated argument of Elliot Sober's new book, *The Nature of Selection*. Why this preoccupation? When even Kenneth Schaffner, its most diehard Columbian defender, found reduction 'peripheral',[35] I really did think the problem had gone away – not because reductionism had triumphed, but because it hadn't. The sciences are too many-faceted to subject themselves to a program of monolithic unification. The new 'tenor' in philosophy of science, as Annis calls it, is happily, hopelessly pluralistic in its import. Not, of course, that anybody is a vitalist, but we recognize the rich and multi-dimensional organization of the natural world, such that the orderlinesses of parts of it not only may be studied in different ways and/or at different levels, but *exist* in different ways and/or at different levels. So, I thought – and I said so in a review paper at the Salzburg congress on philosophy of science in 1983[36] – the reduction issue seemed to be a very dead horse, certainly not in need of further flogging. But here are Kitcher, Burian and Sober all flogging away! What can I say? Two things, perhaps. For one, since Kitcher, Burian, and Sober were trained in the heyday of the 'received view', perhaps they find it hard to forget its problems – some would say, its solutions while ancient individuals like myself lacked that indoctrination in our youth (indeed, I became disillusioned with the 'received view' some years before its official birth – in 1937–8, to be exact!). But there is a more significant point, and that is that in fact as all these writers deal with reduction – even Kitcher, who is dealing with it head on – the old over-abstract question is transformed into a more fruitful one: about the particular *contexts* within which relations between different disciplines arise. The question is always: what needs explaining in the context of this current problem, and what other concepts or principles or methods can appropriately be imported here and now from other disciplines, and whether at a more or less 'micro' level of analysis. As Kitcher demonstrates, molecular biology has indeed, and of course, illuminated, and explained, a great deal that was formerly obscure in 'Mendelian' genetics. But, if, for example, a particular effect is produced by the relative position – the distance apart – of two molecules in the genome, it is the distance, which is spatial, not biochemical, that is causally significant in this case. As Keith Thomson also argues in a recent review of work on development and evolution, there is no one global answer to such questions.[37] It all depends on context – and not only on the context of inquiry, but on the context forced on us by the complex structure of the natural world. Thus in the setting of the newer philosophy of science, another pseudoproblem of the older tradition, with its pseudo-solution, has been assimilated, and translated, into a richer philosophical language, richer both in its epistemological and its ontological import, just because it is frankly cognizant of its historical base. Through its very contextuality, a contextual approach can claim an objectivity that the pretense of total objectivism fatally failed to achieve, and historical realism can claim real rootedness in nature just because it begins where we all begin – and continue and finish – in the real efforts of real historical beings to understand the processes and entities that characterize the world of which they form a real, ongoing part.

Two concluding comments are perhaps in order. First, a word about 'models'. Mayr, I mentioned in passing, thinks concepts more significant than laws in the structure of biological thought. Similarly, 'models' often do more work than 'theories'. In teaching philosophy of biology, for example, I have found especially useful Richard Levins's 1966 paper on the use of models in biology (reprinted by Sober in Part One of his anthology – even though Part Two is, disappointingly, organized wholly in terms of the so-called 'received view').[38] Lisa Lloyd also relies on 'models'; in her usage, however, 'models' are interpreted in terms of that exotic Princeton inflorescence, the so-called 'semantic theory', which in turn accepts an extremely doctrinaire version of the concept of explanation and argues that models, because they are not Hempelian explanations, are not explanatory at all.[39] It would be worth considering how models in Levins' sense fit into the new historical approach to epistemic problems. (I am sorry Sober didn't do that).

Second, I suppose I should, in conclusion, make a bow in the direction of relativism, especially after hearing MacIntyre's Presidential address at the Eastern division of the American Philosophical Association last December.[40] I *think* he was trying something rather like what Annis is doing with his contextualist approach: admitting the multiplicity of contexts of inquiry, that is, of perspectives, and showing how we could manage to achieve epistemic responsibility despite such a plurality. He offered a pair of contrasts of two languages whose speakers were (allegedly) reciprocally without common understanding – the Zuni and the Conquistadores, or the 17th century Irish and English. Then he contrasted these pairs of reciprocally alien languages with a modern critically sophisticated language, possessed of a scholarly tradition, such that one can acquire the language of another group and attempt, imaginatively, to enter, up to a point, into its mode of thought. Scientific discourse is such a 'scholarly' language, both with respect to its own past and to the reporting of extra-scientific experience.

MacIntyre's presentation was dramatic and in a way convincing. What has made the threat of relativism a menace, however, is the presentation of multiple languages – or multiple conceptual frameworks – as necessarily a concession to *ir*rationalism. What we need to recognize, on the contrary, is the reasonableness of each such perspective in its own terms (there may be irrational societies – that's a different, and a difficult, problem – !). Within limits, at least, members of each group assert their fundamental beliefs, as Polanyi put it, *with universal intent* – or in Rowen's language, with due epistemic responsibility. The *relatively* critical stance of a scholar or scientist allows him (her) to concede, and to understand, this kind of responsible perspectivity in the discourse both of his (her) own and of (most?) other societies. There are two further qualifications of MacIntyre's argument that I would have liked to make – and that are needed also as corollaries, or better, presuppositions of the new philosophy of science. I can only just mention them here in conclusion. First, a point MacIntyre did make, but only in passing- that is, that it is not just *languages* that mark out one epistemic perspective from another, but different symbol-systems, taking 'symbol' in the more fundamental sense of symbolic anthropology, in which symbolic activity includes rituals, behavior patterns (techniques), not only the spoken or written word or formula. And the other qualification, which I suggested only parenthetically at the beginning, is the possibility of tying down our historical realism to the realities of nature by grounding it in the Gibsonian ecological approach to perception. But those enterprises I must leave to other occasions and in all likelihood to other persons.

## Notes

1. This paper was written as a lecture for the University of California, Davis, April, 1985. It is a very rough draft of what might – or ought – to be said on this topic.
2. Grene (1983, 1986).
3. Nersessian (1984a), cf. also (1984b).
4. Rowen (1986).
5. Kitcher (1984).
6. Kitcher (1985).
7. Beatty (1982).
8. Depew and Weber (1985).
9. Sober (1984b).
10. Polanyi (1958).
11. Merleau-Ponty (1945).
12. Gibson (1979).
13. Perhaps Rowen has already done so; I have not seen her dissertation, only some papers based on it that she read at Vanderbilt.
14. Rowen (1986).
15. Annis (1978).
16. Ibid. p. 216.
17. I have heard Richard Boyd in a lecture come close to giving an account of this phenomenon, but I am not sure if it exists explicitly in any of his published work. Cf., e.g., Boyd in Leplin (1985) and also quotation from and comment on Whewell in Beatty (1982).
18. Gibson (1979), p. 130.
19. Grene (1974), p. 356.
20. Suppe (1976).
21. Burian in Depew and Weber (1985).
22. Annis (1978), p. 216.
23. Ibid.

24. Nersessian (1984a).
25. Kitcher (1982).
26. Kitcher (1979.
27. That is the role of intensional discourse in the biological sciences. Burian's essay, it should be noted, is in turn the beginning of what will presumably be a book-length study carrying further Carlson's classic work (Carlson 1966).
28. Burian in Depew and Weber (1985), p. 34.
29. Ibid.
30. Ibid.
31. Mayr (1982).
32. Kitcher (1984, 1985).
33. Sober (1984b).
34. Hull (1974).
35. Schaffner (1974).
36. Grene (1985).
37. Thomson (1985).
38. Levins (1966), Sober (1984a).
39. Lloyd (1983).
40. MacIntyre (1985).

## Bibliography

Annis, David (1978), 'A contextualist theory of epistemic justification', *Amer. Phil. Quart.* 15: 213–219.
Beatty, John (1982), 'What's in a word? Coming to terms with the Darwinian revolution', *J. Hist. Biol.* 15: 215–239.
Carlson, E.A. (1966), *The Gene: A Critical History* (Philidelphia and London: W.B. Saunders Co.).
Depew, D.J. and Weber B.H. eds. (1985), *Evolution at the Crossroads: The New Biology and the New Philosophy of Science* (Cambridge, MA.: Bradford Books).
Gibson, J.J. (1979), *The Ecological Approach to Visual Perception* (Boston: Houghton – Mifflin).
Grene, M. (1974), *The Understanding of Nature* (Dordrecht: Reidel).
Grene, M. (1983). 'Empiricism and the philosophy of science, or *n* dogmas of empiricism', in *Epistemology, Methodology, and the Social Sciences,* R.S. Cohen and M. Wartofsky, eds. (Dordrecht: Reidel), pp. 89–106.
Grene, M. (1986), 'Philosophy of biology, 1983: Problems and prospects', *Proc. 8th Int. Cong. Logic, Meth., Phil. Sci.,* VII, pp. 433–452.
Grene, M. (1986), 'Perception and interpretation in the sciences: Toward a new philosophy of science', forthcoming in the *Bordet Lectures,* C. Knight, ed. (Belfast: Queen's University Press).*
Hull, D. (1974), *Philosophy of Biological Science* (Englewood Cliffs, N.J.: Prentice Hall).
Kitcher, P. (1978), 'Theories, theorists, and theoretical change', *Phil. Rev.* 87: 519–547.
Kitcher, P. (1982), 'Genes', *Brit. J. Phil. Sci.* 33: 337–339.
Kitcher, P. (1984), '1953 and all that: A tale of two sciences', *Phil. Rev.* 93: 335–373.
Kitcher, P. (1985), 'Darwin's achievement', in *Reason and Rationality in Natural Science,* N. Rescher, ed., (Lanham, MD.: University Press of America).
Leplin, J., ed. (1984). *Scientific Realism* (Berkeley: U. of California Press).
Levins, R. (1966). 'The strategy of model building in evolutionary biology', *Amer. Scientist* 54: 421–431.
Lloyd, E. (1984), 'A semantic approach to the structure of population genetics', *Phil. Sci.* 51: 242–264.

* *Note:* This essay was rewritten into unintelligibility by the copy-editor of the Depew and Weber volume; interested readers are asked to wait for a more accurate version in the Bordet collection.

MacIntyre, A.C. (1985), 'Relativism, power, and philosophy', Pres. Addr., East. Div. APA, Dec 1984, APA *Proc. and Add.* 59: 5–22.

Mayr, E. (1982), *The Growth of Biological Thought* (Cambridge, MA.: Harvard University Press).

Merleau-Ponty, M. (1945), *La Phénoménologie de la Perception* (Paris: Gallimard).

Nersessian, N. (1984a), 'Aether/or: The creation of scientific concepts', *Stud. Hist. Phil. Sci.* 15: 175–212.

Nersessian, N. (1984b). *Faraday to Einstein: Constructing Meaning in Scientific Theories* (Dordrecht: Martinus Nijhoff).

Polanyi, M. (1958). *Personal Knowledge* (London: Routledge and Chicago: University of Chicago Press).

Rowen, L. (1986). *Revisioning Epistemology: A Reflection on Truth, Risk, and Epistemic Responsibility in Scientific Practice*, forthcoming Vanderbilt Dissertation.

Schaffner, K. (1974), 'The peripherality of reductionism in the development of molecular biology', *J. Hist. Biol.* 7: 111–139.

Sober, E., ed. (1984a). *Conceptual Issues in Evolutionary Biology* (Cambridge, MA: Bradford Books).

Sober, E. (1984b). *The Nature of Selection.* (Cambridge, MA: Bradford Books).

Suppe, F. ed. (1976). *The Structure of Scientific Theories*, 2nd. ed. (Urbana: University of Illinois Press).

Thomson, K. (1985). (Essay review: the relationship between development and evolution? *Oxford Surveys in Evolutionary Biology,* 2, R. Dawkins and M. Ridley, eds. (Oxford: Oxford U. Press).

# Research Problems and the Understanding of Science

*Reston, Virginia*

Most of us learn of science first through text books, and these as a rule present their subjects – electricity, heat, etc. – as complete and empirically verified systems. This is not done dogmatically, in that open problems are sometimes noted in passing, or limitations are implicitly admitted by labelling their subject matter as 'Classical' or as 'Introductory'. The students who go on to research find themselves immersed in quite a different world. They are first of all required to find a research problem – a place where science is incomplete or unclear – and they become aware that there are, in their field, contending programs for research – different ideas about what is important and what direction research should take. Those things which, in the text-books, seemed on the periphery of science now become central to it.

One of the developments in history and philosophy of science over the past twenty-five years has been to include research problems and programs as an impor-tant part of our picture of science. This development constitutes a welcome advance over traditional views, which picture scientific theories as emerging purely from 'the facts', or from the facts plus manipulation of formalism. Better understanding of research problems, in particular, can be of great benefit to both scientific researchers and historians. For the researcher, a knowledge a the problem a theory or experi-ment was originally intended to solve can provide a valuable insight into the strenghts and weaknesses of current theory and data. For the historian, identifica-tion of the problem that the innovation was a response to does remove a great deal of the mystery about the innovation – though of course a good deal remains. (The most informative theories of creative thinking – the Gestalt, Würtzburg, and Artificial Intelligence theories – are all theories of problem-solving.) Finally, for both histo-rians and scientists a knowledge of the original problems is particularly valuable in understanding scientific concepts which are far from those of common sense, such as the concepts of modern physics; for historically the evolution has been from common sense notions to those which, under the pressure of the need to solve problems, have become further and further removed from common sense.

Though scientific problems have been given a more prominent role in recent theories of the growth of science, their importance as a driving force in research has not been fully recognized. In the most widely read works on research programs,

*Nancy J. Nersessian (editor), The Process of Science. ISBN 90-247-3425-8.*
© *1987, Martinus Nijhoff Publishers (Kluwer Academic Publishers), Dordrecht. Printed in the Netherlands.*

those of Thomas Kuhn and Imre Lakatos,[1] problems are seen as derivative from a particular research program. For Kuhn, the scientist's normal task is to solve 'puzzles' by applying the 'paradigm'. For Lakatos, the scientists attempts to predict novel facts to confirm the 'hard core' of a 'scientific research program'. In both cases the problems are hurdles to overcome in carrying out a single research program, and in that way subordinate to it.

In reality, a research problem is not simply an appendage of a particular research program. Rather, research problems and programs interact with one another. It is true that particular research problems are often chosen because they potentially further a particular research program. And furthermore, the research program forms an important part of the *problem situation* – the constraints on what kind of a solution the research is looking for. However, this is by no means the only reason a research problem is chosen. A scientific research problem may be of interest because it promises to shed light on a *clash* between different world views. Or it may be worked on because it solution promises to be of help to medicine or to technology. Furthermore, a problem, once selected for research, has a certain flinty independence. It sometimes refuses to bend to a research program, and the scientist will change or abandon the program in order to solve the problem.

Perhaps the best way I can make clear the case for the greater importance of problems in the understanding of science is by a look at some of the recent history of the problem of scientific problems itself. Interest in the role of problems in science is itself the result of attempts to solve the problem of *fundamental change* in scientific theories. In the traditional view of science, scientific theories are generalizations of observation, and scientific knowledge grows *cumulatively* by discovery of new facts, the addition of new generalizations, and the extension of old ones. The fundamental changes in physics in the first part of this century – relativity and quantum theory – shattered this cumulative view.

The contradiction between what was happening in physics and the traditional view created two related problems. The first was a problem of *standards:* if the new theory was not an extension of old ones, but a replacement, then how could we judge it as any better? After all, the new theory might itself be overthrown. The second was a problem of *creation:* if theories are not created by generalization and extension, then how are they created?

Some thinkers, beginning with Poincaré, tried to argue that the development of physics, properly understood, did not involve overthrow and replacement, so that the problem was not really a problem. The first one to face the problem squarely was, I believe, Karl Popper in his first book *The Logic of Scientific Discovery.*[2] And a distinctive view of scientific problems was a key part of his solution.

Popper brought to the problem of scientific change much of the outlook of the German cognitive psychology he had studied in the twenties.[3] According to O. Külpe and his followers, the Gestalt and Würtzburg schools of psychology, problems are a basic motivation to creative thinking. Independently of Külpe, C. S. Peirce had also argued for a problem-based view of creative thinking, and following him John Dewey (with whom Popper was familiar) advocated similar views. All of these thinkers were under the influence of Darwin, and any view which holds that trial and error is a central aspect of the growth of knowledge naturally includes problems as a stimulus to trial and error.

Popper had studied with Bühler, a leading contributor to the Würtzburg school, but, inspired by Einstein's revolution in gravitational theory, he brought his own ideas about problems to account for the growth of science. In Popper's view, scientific problems are largely of the character of *contradictions* between freely posited theories on the one hand, and incompatable observations or experiments on the other. Instead of science growing by a process of generalization, of induction from lower levels to higher, science grows by a process of 'conjectures and refutations', in which all levels of generality are simultaneously involved. Tentative solutions to problems are, in his view, explanatory theories conjectured to explain the recalcitrant data. Furthermore, experiments are designed as *tests* of theories, rather than being simple efforts to catalogue facts as a preliminary step before generalization.

Popper used this idea of problems as contradictions between universal theories and particular facts to solve the problem of radical scientific change. In the matter of standards, the explanatory power of the new theory could be a measure of progress, even when overthrow of past theories took place. The new theory, if it were to be a decisive improvement, would have to explain (by a formal deduction) the facts the old theory explained, and, preferably, more besides. Since the old theory did have some success, there would be an analogue or approximation of the old theory constructed within the new, and this would also account for the success of the old theory. And thus we could measure progress even while accepting that genuine refutation and overthrow of theories take place.

On the aspect of creation of new theories and experiments, we could also account for these by other than generalization. The theories are attempts to resolve the contradictions between theory and experiment – a kind of trial and error problem-solving. And new observation and experiment is an attempt to test existing theory to find its strengths and weaknesses. Popper himself did not emphasize the creative aspect of change in his first book. He did return to it later, however, emphasizing the role of problems in all creative thinking.

Whereas Popper had focussed on the problem of standards during radical change, two thinkers accepting Popper's view that there are genuine revolutions focussed on the problem of creation, and, in the late fifties and early sixties produced a richer picture of the growth of science. These are Joseph Agassi and Thomas Kuhn[4] and both focussed on research programs as a guide and inspiration to scientific creativity.

In Kuhn's view, most science is not of the 'revolutionary' sort Popper focussed on, but 'normal science', which prepares the way for eventual revolution. Normally, scientists take the work of a great revolutionary, such as Galileo or Einstein, as an exemplar or 'paradigm' and try to use the same methods to solve other scientific problems. These problems are, however, of a different character than Popper had described. They are not contradictions, but what Kuhn calls 'puzzles': the problem is, in effect, to apply past theory and technique to explain heretofore unexplained phenomena. Only when sufficient recalcitrant cases or 'anomalies' have built up is a field ripe for revolution, when a new scientific genius can come along and create a new 'paradigm'.

Agassi presented a quite different picture of the function of research programs. He accepted Popper's characterization of scientific problems, but pointed out that the *selection* of problems to work on has been influenced strongly by issues in metaphysics. In particular, research has centered on issues of *competition* between differing world views, such as the conflict between the atoms and empty space view and the field view of matter and its interaction with other matter. The metaphysical world views are not in themselves testable, being claims such as that the world can be explained in terms of atoms or fields. However they do function as research programs, and help guide not only the selection, but also the solution of scientific problems. In Agassi's view, scientists typically look for experiments and theories which admit of a ready explanation or formulation in terms of one world view, but not te other. In other words, the scientist attempts to crystallize the world view into a testable scientific theory which can not formulated in the other world view. In this way, science produces concrete and powerful arguments for one world view and against another. To understand the growth of science, then, we need to view it within the context of a larger debate over the correct world view, a debate which includes considerations beyond scientific ones.

In the late '60s, Imre Lakatos[5] became dissatisfied with Popper's solution to the problem of scientific standards during radical change, and tried to improve on it by synthesis of Kuhn's and Agassi's views. Lakatos saw that Agassi's view of *competing* research programs was more realistic, but felt also that Kuhn's picture of people out to confirm research programs – rather than test theories – was the key to a better view of standards. According to Lakatos, scientists set out to confirm the 'hard core' of a 'scientific research program'. The way they do this is by adding additional assumptions to the 'hard core' of assumptions in order to predict novel facts. A research program is not in isolation, but has competitors with a different 'hard core' of assumptions. The research program that can keep predicting new novel facts is 'progressive', whereas one that can merely adjust to new information after the fact is 'degenerating'. Continually progressing research programs are chosen above degenerating ones – and should be so chosen. The standard for judging the progress of science in radical change is not, then, the *theory* which passes tests, but rather the *research program* which is most progressive.

Since Lakatos's widely read effort, there have been many discussions of research programs. The most insightful I have seen is Jagdish Hattiangadi's 'The Structure of Problems'.[6] Like Lakatos, Hattiangadi was initially concerned with the problem of standards; as he puts it, he was concerned with why scientists should be so concerned to replace one false theory with another – which is, in Popper's view, what is happening in the growth of science. Hattiangadi was dissatisfied with Popper own arguments regarding the increasing 'verisimilitude', and proposed another solution. We propose new theories because they solve problems the old ones did not, and their superiority lies in this power to solve problems. However, Hattiangadi felt that the problems involved were not simply the contradictions between universal theory and specific fact; i.e. the ones emphasized by Popper. Rather they should include also the conflicting traditions in metaphysics which Agassi has pointed to.

Hattiangadi, then, proposed a broadening of Popper's notion of a problem: all intellectual problems may be thought of as *contradictions,* and a solution as a resolution of the contradiction. The new solution may give rise to new problems – new contradictions – and Hattiangadi goes to some lengths to show that in spite of this there is a clear standard for the new solution being superior to an old one. In fact, as Hattiangadi shows, one can identify a complex structure of problems (contradictions) and solutions developing through history. This analysis is somewhat reminiscent of Hegel's dialectics, but since the notion of 'contradiction' is taken from modern formal logic, and is quite different from Hegel's, it has a clarity and precision lacking in Hegel's historical methods.[7]

One of the valuable features of Hattiangadi's analysis of problems is that he gives us a series of sharply formulated questions to penetrate intellectual history and gain an understanding of it: What contradiction was the particular scientist trying to resolve? What was the proposed solution? The questions apply not only to the type of problem Popper focussed on, the contradiction between a theory and observation, but also the contradictions between different metaphysical world views, such as Agassi had pointed out. These questions thus are a tool for identifying not only the immediate problem the scientist faced, but also the whole intellectual tradition which led up to that problem, including rival research programs.

Hattiangadi's analysis also makes clear the fundamental error in Kuhn's and in Lakatos' view of the role of problems in science. Agassi had pointed out that the key problems which led to scientific revolutions were problems involving the *conflict* between different metaphysical research programs. Hattiangadi's analysis of the structure of problems makes clear that problems acquire their significance precisely because they express the *clash* of ideas. In Kuhn, the clash of ideas is not important: science is normally the activity of a person trying to apply a paradigm in explaining various phenomena, and not to resolve a clash of views. For Lakatos the clash of research programs is an important part of science, but the actual problems which the scientist chooses to work on are determined within the research program, and not by the clash of ideas.

As Hattiangadi aptly puts it, Kuhn and Lakatos have a kind of 'impetus' theory of the growth of science. Scientists adopt a 'paradigm' or 'research program' and go forth to conquer with it. As they succeed the research program keeps its impetus. As it fails, it 'runs out of gas', or of impetus, and either is replaced by an existing more successful rival (Lakatos) or by a new revolutionary theory (Kuhn). This 'impetus' view of science has, in my view, a very undesirable element of conservatism in it, as it makes criticism of an established 'paradigm' or 'research program' irrelevant and possibly harmful to science.[8] In the 'clash of ideas' views of Popper, Agassi and Hattiangadi, such criticism might prove an invaluable spur to the growth of science. Furthermore, the 'impetus' view is not, in reality consistent with the facts of the history of science. Not only were there rival research programs, but the points of clash were of critical interest to the revolutionary scientists.

In spite of the considerable light Hattiangadi's analysis sheds on the growth of science, I do not think his complete identification of scientific problems with kinds of contradiction holds up to scrutiny. In the first place, there are scientific problems generated by other means than a conflict between different world views, theories, or theories and evidence. These are the scientific problems which are taken up because their solution will, it is hoped, aid the solution of practical problems concerning technology and medicine. For example the problem of explaining why cancers spread does not arise because cancer violates some theory, but because of the search for a cure, and the belief that better understanding of the cause will aid the discovery of a cure. However, the problem of understanding the cause of the spread is a genuine scientific problem, not simply a practical one. A practical problem is one whose solution is an action, or decision on a course of action. But the problem of finding the cause of cancer's spread may be solved fully without being able to solve the practical problem of preventing the spread.

The bulk of science done today is, in fact, of the 'applied science' type. It applies existing fundamental theories to the solution of scientific problems which, it is hoped – often correctly – will aid in the solution of practical problems in medicine, industry, weaponry and so on. These problems of applied science do have the structure of 'normal science', as described by Kuhn; Kuhn has put his finger on a dominant kind of science which Popper overlooked. But, ironically, it is *not* this kind of science, applied science, which has been responsible for the major scientific revolutions. It is rather the kind involving the clash of world views discussed by Agassi and Hattiangadi.

A second reason for rejecting Hattiangadi's identification of scientific problems with contradictions is that the contradictions only arise when a particular kind of explanation is sought. Let me clarify this point with an example outside science. In the ancient world, it was common that a conquering people would impose its theology on the conquered people. However, this was not usually a big problem to the conquered people; for in polytheistic systems there are often Gods with similar functions – a God of rain, of fertility, of destruction, and so on. Thus a people could rename their old Gods, combine the new stories of the Gods with the old, and so on. There was no sharp contradiction between one polytheistic theology and another.

However, there was a contradiction between the polytheistic systems and the ethical monotheism of the Jews and subsequently of the Christians. The basic idea (of Abraham?) was that there are not many Gods who take different people's side, but one God who judges people on the basis of the justness of their actions. This idea of one God of justice could not be harmonized with a polytheistic system, and the Jews and Christians refused to adopt the pantheon of Gods of the Romans who ruled them.

The kind of explanations which scientists seek have this unified, comprehensive character – like monotheism – so that the different world views *can* contradict one another. The proto-scientific theories of the early Greeks – all matter is a form of water, all matter consists of atoms and the void, etc. – all have this character. The most accurate characterization of the kind of explanations scientists have sought is I think that of Émil Meyerson.[9] According to Meyerson, our search for understanding generally is a search for *identity* within diversity and change. This search for identity in science takes two particular forms: first the attempt to discover *things* – atoms, genes – which *persist* through change; and, second, the attempt to find unchanging *laws* which govern the changes that do take place. I think that in order to characterize problems in *modern* science we need to add, with Popper, the requirement that the description of things and laws needs to be *testable* by observation and experiment. If instead of looking for this kind of explanation, scientists would explain the stormy sea by saying 'Neptune is angry', and so on, the contradictions that Hattiangadi points out would never arise in the first place.

Problems – intellectual or practical – can be characterized as a difficulty in achieving an aim.[10] Scientific problems I think are best characterized as difficulties in achieving the aim of *testable explanations* of the world of our experience – where 'explanation' is meant in Meyerson's sense and 'testable' in Popper's. It is true that contradictions between world views are a particularly important difficulty in explanation, and that they have been the focus of the efforts that have led to scientific revolutions. However, they are not the only kind of scientific problem. The problem of why the paint peels on my front gate is a scientific problem, though it may be of no special interest to scientists working today. It can acquire interest either for practical reasons – a paint manufacturer wants to produce longer lasting paint – or because it is the focus of a clash in world views – say two different views of the nature of decay, as in the second law of thermodynamics. It can also acquire importance as a test – a potential clash between theory and experiment.

The ideas of Agassi, Kuhn, Lakatos and Hattiangadi, then, can be seen as theories of the way *problem situations* evolve and change in science – where by 'problem situation' I mean the circumstances that lead a scientist to select a problem a worthy of effort, and that guide him in its solution. As such, I think all of them have valuable insights into the creative process in science. However, the scientific problem should not be absorbed into problem situations (as Hattiangadi attempts), nor research programs nor 'paradigms'. Scientific problems have a certain flinty autonomy, and refuse to be bent into shape to fit any particular research program. For example,

Planck is said to have introduced the idea of the quantum jump not because he liked it, but because it solved the problem he faced. And Einstein, in explaining the photo-electric effect by a more radical version of quantum theory, produced a theory he was sufficiently dissatisfied with (and outside his research program of relativity) to call it 'heuristic' – but it solved the problem. This kind of stubbornness of scientific problems I think reveals the element of truth in P.W. Bridgeman's statement that the scientific method is to 'do your damnest, no holds barred'. The question is, do your damnest to do what? And the answer is: to solve the problem. When we clearly identify that problem – as well as the situation as the scientist viewed it – then we understand the scientists' activity in a way that no other piece of information will give us.

Though Kuhn and Lakatos do not, I think, get the role of problems quite right in the development of science, they do give problems an very important role in the development of science. Indeed, all those philosophers discussing research pro-grams do give problems an important role; the research program or paradigm in one way or another guides the selection and solution of research problems. Many historians of science less concerned with the general issue of creativity in the sciences have been more hesitant in dealing with scientific problems. While particular research problems have sometimes been identified, there has been a great reluctance to recognize the investigation of research problems and problem-situations as a *systematic* method of research in intellectual history. And the reason lies, I think, in something that Popper was the first to recognize: a problem-based view of the creative process is, in fact, incompatible with inductive proof as the basis of science. The main appeal of the inductive view is that it supports the idea that scientific theories, once derived from data, are reliable and will not be overthrown. And the same factor appeals to historians of science in their work: if they have derived their view from the facts, then it will be correct and reliable.

However, once we acknowledge that there is an element of conjecture in our interpretation of history, then the lasting value of our work as historians is threat-ened. But, if the problem-based view is correct, the element of conjecture is unavoidable.[11] Popper's answer to this difficulty is that testing our conjectures against the evidence provides a way of weeding out false conjectures. Any view of reality, whether of natural law or of history, which is able to pass such tests can hold up its head proudly, whether or not it is eventually in need of modification.

The historical method I advocate, then, is to conjecture the problem and problem-situation as the scientist understood it, and then check this conjecture carefully against the written evidence we have. The effort to avoid conjecture either im-poverishes our history by leaving out the dimension of problems, or it allows our conjectures to slip into our account unawares, without testing against the evidence they should be compared with. We should, rather, acknowledge the conjectural character of our attempts to reconstruct the problem but make the attempt, and check it against the evidence.

The positive reason to give problems a central place in our histories is that problem-solving views of creative thinking are the only viable models of creative thinking that we really have. For while it is true that these theories (of Peirce, the Gestalt School, the Würtzburg School, Popper, and of the contemporary researchers into Artificial Intelligence) are rather rudimentary, we have no rival to them which gives us any guidance in reconstructing and understanding the process of discovery.

An example will, I think, make clear the necessity of including problems as a part of insightful history science. In my book *Fields of Force*[12] I tried to reconstruct the problem situation of each major figure in the development of field theory from Faraday to Special Relativity: Faraday, Maxwell, Helmholtz, Hertz, Lorentz and Einstein. One bit of the history which has stirred some controversy is the question of when Faraday developed his field theory. In my reconstruction of his problem situation, and his efforts to solve the problems he perceived, I felt that his basic theories where pretty fully developed by the time he finished his second series of experiments discovering electromagnetic induction. His later experiments in electrostatics, electrochemistry and so on I view as attempts to find a crucial experiment which would confirm his field views and refute the action-at-a-distance view.

L. Pearce Williams[13] has taken issue with this, arguing that his ideas where only developed later, after many other experiments. And D. Gooding[14] seems to concurr. N. Nersessian[15] has taken a third position, arguing for a gradual development of Faraday's theory over the course of his researches. Agassi[16] puts the date of his field theory even earlier than I, at the time of his discovery of the principle of the electric motor, in 1821. I do not wish to argue the case here, as this would be out of place, but rather to point out that there is no way to assess who is right except by conjecturing what problems Faraday was trying to solve, and what ideas he had on how to solve them. And then, of course, we need to see how these conjectures stack up against the evidence.

In Faraday's case this conjecturing is clearly necessary; for the 'inductive' view of the development of his ideas – that his ideas emerged purely from his experiments – is refuted by the documentary evidence. In 1816, when he was twenty–five, Faraday said that he believed that chemical affinity and electrical attraction are the same. In 1821 he gave arguments as to why we should expect an electrical effect of magnetism. Over the next decade he did several series of experiments to find such an effect, and did find it – electromagnetic induction – in 1831. In the papers describing the discovery of electromagnetic induction he adumbrated his theory of electrochemistry, but cut it short 'though much tempted' to offer 'further speculations'. He immediately began his researches in electrochemistry, again making important discoveries, but also at the same time sent a sealed note to the Royal Society to establish priority for the idea that electrical and magnetic actions have a finite velocity. This idea he only published years later, after his discovery that magnetism could rotate the plane of polarization of a ray of light.

The record is thus littered with examples – and these are only a few – of Faraday's hints and speculations *before* he made his discoveries. Of course, the results of the experiments did affect his thinking, but the evidence is overwhelming that vigorous creative thinking going beyond the evidence led him to devise the experiments. Why was Faraday speculating? How were these speculations instrumental in his discoveries? How did his discoveries modify his speculations? My own view is that one of Faraday's fundamental problems was to find a unified theory of electricity and the chemical bond. And when he turned to problems of electricity and magnetism, he never left this problem behind: whatever theory of electromagnetism he could devise, it must also have the potential of explaining the chemical bond. In reconstructing Faraday's thinking, the thinking which led to the experiments and followed from them, we have two touchstones, so to speak. On one side we have the problems: given what we know of the problems that Faraday was thinking about (such as the chemical bond), do the ideas we conjecture he had at that stage make sense as solutions? Secondly, are the experiments as reported actually *tests* of those ideas – that is, attempts to confirm or refute those ideas, particularly in relation to competing views? We have in addition, of course, both Faraday's earlier hints on his thinking, and the later, most fully expressed ideas.

Those taking issue with one or another aspect of my reconstruction of Faraday's thinking have not, thus far, pointed out either any inconsistency with the evidence, nor with the requirement that the ideas make sense as a solution to the problems as Faraday understood them. Thus the debate, in these terms, has not really been engaged. The general point I wish to make is that, as in the case of Faraday, to reconstruct the process of discovery in science conjecture is necessary, and the identification of he particular problems the scientist attempted to solve is essential. For only then can we make sense of the historical record. And thus both for the understanding of the creative process in science generally, and for the understanding of specific historical incidents, the identification of scientific problems is of vital importance.

### Notes

1. Kuhn (1962); Lakatos (1970).
2. Popper (1934).
3. For the background of Popper's ideas in German cognitive psychology see Berkson and Wettersten (1984).
4. Agassi (1964); Kuhn (1962).
5. Lakatos (1970).
6. Hattiangadi (1978).
7. c.f. Popper's essay 'What is Dialectic', in Popper (1962).
8. For criticism of Kuhn's conservatism, see Berkson (1974a).
9. Meyerson (1930).

10. Hattiangadi does characterize a problem as a difficulty in achieving a goal, but then drops discussion of the goals of science, and only discusses the difficulty of contradictions. He attempts to derive Popper's demand for independent testability from his own theory of the structure of problems, but the derivation seems to me to be illegitimately circular. An indication of the need of specifying an aim of science is the fact that Hattiangadi's characterization of conflicting traditions, debates, and tests could equally well apply to, say, the efforts to correctly interpret a dialogue of Plato. Yet such a problem is not a problem of science. I suspect that if we include the aim of testable explanation, some of the logical difficulties Hattiangadi discusses could more easily be resolved.

11. See the extensive study of the implications of different philosophies of science for the history of science in Agassi (1963).

12. Berkson (1974).

13. Williams (1975).

14. Gooding (1976).

15. Nersessian (1984).

16. Agassi (1971).

## Bibliography

Agassi J. (1963), *Towards an Historiography of Science,* Beiheft 2, *History and Theory.*

Agassi J. (1964), *'On the Nature of Scientific Problems and their Roots in Metaphysics',* in The Critical Approach, ed. Mario Bunge (Free Press, New York); reprinted in Agassi's Science in Flux (Reidel, Dordrecht 1975).

Agassi J. (1971), *Faraday as a Natural Philosopher* (Chicago U. Press, Chicago).

Berkson W. (1974), *Fields of Force: The Development of a World View from Faraday to Einstein.* (Routledge & Kegan Paul, London).

Berkson W. (1974a), 'Some Practical Issues in the Recent Controversy on the Nature of Scientific Revolutions' in *Boston Studies in the Philosophy of Science, 14,* R.S. Cohen and M. Wartofsky, eds (Reidel, Dordrecht).

Berkson W. and Wettersten (1984), *Learning from Error: Karl Popper's Psychology of Learning* (Open Court, La Salle, Il.)

Gooding (1976), Review of *Fields of Force,* in *British Journal for the History of Science,* p. 89–91.

Hattiangadi J. (1978), 'The Structure of Problems', in *Philosophy of the Social Sciences 8,* 345–65; *9,* 49–76.

Kuhn T.S. (1962), *The Structure of Scientific Revolutions* (Chicago U. Press, Chicago).

Lakatos I. (1970), 'Falsification and the Methodology of Scientific Research Programs', in *Criticism and the Growth of Knowledge,* I. Lakatos and A. Musgrave, eds. (Cambridge U. Press, Cambridge).

Meyerson (1930), *Identity and Reality,* English translation 1930, Dover Reprint 1962 (New York).

Nersessian N.J. (1984), *Faraday to Einstein: Constructing Meaning in Scientific Theories* (Martinus Nijhoff, Dordrecht).

Popper K.R. (1934), *The Logic of Scientific Discovery,* English translation 1959 (Basic Books, New York and Hutchinson, London).

Popper K.R. (1962), *Conjectures and Refutations* (Basic Books, New York).

Williams (1975), Review of *Fields of Force,* in *British Journal for the Philosophy of Science 26:* 241–53. And my reply and Williams' further comment in the same Journal *29:* 243–252.

# Twenty Years After

JOSEPH AGASSI

*Tel-Aviv University. York University, Toronto*

One of the greatest monuments to the human intellect is the Babylonian Talmud, compiled, we are told, about the year 500. In it the two chief compilers congratulate themselves, with understandable pride, saying, we are no mere woodcutters in the swamp. The chief compiler is characterized elsewhere in that vast work as a master dialectician – evidently with his consent. He was, it is said, one who would raise a whole palm-tree and then take an axe to it. The question, what good does it do put all the labor required into the act, what is gained by raising a palm-tree and then cutting it down, this question could not possibly occur to him. He knew that preoccupation with the law is the highest form of activity, the best form of life. The idea of progress was alien to him, if not actually distasteful. Echoing Plato, the Talmud says, if the former generations were angels we, the later ones, are merely human; if the former were human, we the later ones are asses. Yet, clearly, these humans or asses saw in the corpus that was compiled an achievement. How does one reconcile the idea of progress with a static system of thought? I do not know. I do not think it possible.

Let me stress, the question is not, is the compilation of this or that scholarly work progress? The question is, does a certain philosophy permit progress at all? This question is a bit vague. Clearly, the Talmudists denied the possibility of progress in some sense yet affirmed the possibility of progress in another sense. Moreover, both views are stated quite emphatically. Nor is there a need to declare that a contradiction is present in this. Clearly, before one could judge matters one needs to be told in what sense the Talmudists saw progress as impossible and in what sense they saw progress as possible and even present. Yet it is not easy to say; at least I find it hard to say. Clearly, in the sense in which I can elucidate matters somehow, I may employ the jargon popular today among both philosophers of science and historians of science, a jargon usually ascribed to Thomas S. Kuhn. But I wish to stress that I am using here a popular jargon, not Kuhn's own views. I have heard Kuhn often say he is misrepresented, and I would rather not represent him at all just because of that recurrent complaint of his. And I also wish to stress that I am using the jargon loosely since it has but a loose meaning – which fully accords with Kuhn's recurrent complaint.

*Nancy J. Nersessian (editor), The Process of Science. ISBN 90-247-3425-8.*

Well, then. The Talmudic scholars were operating within a paradigm. They denied that progress is possible in the sense that they deemed the paradigm absolute, contrary to contemporary views of paradigms as changing from epoch to epoch. The reason the Talmudists saw the paradigm as unalterable is, of course, their attribution of it to divine origin. They felt, quite rightly in my opinion, that without such divine attribution, adherence to a paradigm becomes a frivolous affair. One may, of course, adhere to a paradigm for pragmatic purposes. This possibility is, historically, the most popular second option. The two options, incidentally, are known in the philosophic literature by a number of catch-phrases, such as nature and convention, absolutism and relativism, dogmatism and pragmatism.

What is common to absolutism and relativism is the idea that progress remains within a paradigm. What they differ about is the number of paradigms available, whether it is one or many, tu use another catch-phrase.

As historians we may wonder, what offers a historian more scope, the one or the many. The attraction of the many is what comes forth first: many paradigms present many epochs: from the very start the historical dimension is introduced. Yet this is the hisorian's view. The philosopher's view is quite the contrary: the one is the resting place, the home our spirit yearns for; the many only disturb us by placing before us an insoluble problem of choice: which of the many should I choose? The historian does not choose. For the historian, as a fact, Tom was a mechanist and Dick a vitalist, Reuben a plenist and Simon a vacuist. It is given; it is no problem of choice.

Some great thinkers have tried to make use of this fact, to approach philosophy historically. Prominent among them is Martin Buber who claimed that the choice of a religious paradigm is a matter of historical fact. Following him Michael Polanyi declared that the choice of a scientific paradigm is also a matter of historical fact. Polanyi did not mention Buber, as far as I know, but he used Buber's unmistakable idiom. Anyway, their idea is that we can choose only those conventions or paradigms which exist as cultures or as cultural communities. Moreover, they said, you need not like any existing option, since you can alter it, but you can alter it only from within, i.e., by endorsing its paradigm to begin with.

When comparing the Talmudic scholars with Martin Buber, we may see a remarkable shift in the very concept of the paradigm itself. For, the Talmudists declared as central an unchangeable core doctrine, the divine word, and change and progress as occurring only in its application, interpretation, and further application. For Buber, and for Polanyi, as long as the socio-cultural base maintains continuity, all is well; the doctrinal changes are matters of the community in question, and no outside philosopher, certainly not a Buber or a Polanyi, has the authority to dictate any rule on that matter.

This is, briefly, a cop-out. The social background is considered primary and the culture it carries is used as a mere excuse for our study of it, or a rationale for our study. And this erroneous order of priorities limits one's very ability to understand the culture whose social background one carefully analyzes. For, living societies face problems, problems rooted in the social background but not soluble by looking at that background. Problems beset any active member of any living community, be that community the Catholic Church, the local Jewish synagogue or the International Society for the History of Science. The problems are varied. Some are soluble within the paradigm in a reasonable manner, some not; is reform then the order of the day? Neither Buber nor Polanyi has an answer. Neither has a criterion. They even explicitly declare that no criterion exists but that the leadership, the spiritual and intellectual leadership, take responsibility and act with no criterion to guide them. Polanyi knowns that leaders can err. They may then lose their position at the helm or, much worse, they might take the whole ship to a reef. All this is true, but quite unhelpful to the leaders. The leaders are not averse to being offered advice. They are, obviously, averse to having advice rammed down their throat, but, equally obviously, not amused when advised to act without the aid of advice.

Perhaps I am too rash to say they are willing to listen to advice before they make up their own mind. As a budding philosopher studying science and moving in the company of normal scientists I was told many a time, and in a blunt manner not meant to spare my feelings, that philosophers of science are individuals trading in advice that is both unasked for and inherently useless. But what they meant thereby to affirm, was their status as normal scientists, as individuals who take the paradigm of their scientific community as given, and as individuals happy to perform tasks which are valued by their community, no questions asked.

There is something terribly important shared by the ancient Talmudic scholars and today's scientists. Something which I consider the root of their peace with themselves and profound inner sense of well-being. It is the obviousness of their choice. The individuals in question know that they had a choise of a paradigm. Many Talmudic scholars were steeped in the local culture within which they lived in exile, yet they never stopped to consider their choice: the moral and intellectual superiority of their Jewish paradigm over the local culture hardly invited articulation. The same holds for my scientific friends. They could become religious thinkers or otherwise alter profession, but they knew they had a calling.

Yet, I must repeat, all this holds for the rank-and-file, for the normal scientist or the normal Talmudist. The leading ones faced more basic problems and sought criteria by which to face them. One of the very earliest Talmudist leaders, Raban Gamliel, decided to rely on scientifically calculated calendars in preference to traditional ways of determining dates by observing the new moon. He was challenged. He used his authority and demanded of his challenger to publicly accept his own – scientific – ruling, in preference to the challenger's traditional alternative. How did Raban Gamliel know that the heathen scientific calendar is reliable? How did St. Robert Cardinal Bellarmine know that the Copernican calendar is more reliable than the Julian calendar instituted by Raban Gamliel? (For, the Gregorian calendar is but a variant, well within the Julian paradigm.)

These are the obvious questions to ask anyone who is not quite content to be a normal member of a given intellectual community. Thomas S. Kuhn is right about one point in this context: he notices that at times this question is not pressing, at times it is. And when the pressing need for a change is felt and the leadership successfully effects the change, then the sooner the rank-and-file endorse it, the more easily they can pretend that nothing much has happened. That is to say, Kuhn views paradigm shifts from the viewpoint of the obedient rank-and-file. For the obedient rank-and-file are content to leave the big tough questions to others and merely follow in their wake. They want no advice from outsiders, they want instructions from the leaders. But not all rank-and-file are as obedient as he describes. He wrote about the Copernican revolution and he wrote about the quantum revolution and in neither case did he notice the enormous qualms and perplexities which even simple and very normal researchers may suffer. When you are perplexed, says the great Guide to the Perplexed, go and consult an acknowledged wise person. How do I know whether the acknowledged wise is wise? Once you ask this question, you are obedient normal rank-and-file no longer. The obedient normal rank-and-file have no trouble knowing this. You go to a convention and snoop around. In no time you find out. I think this is true. I think Maimonides' *Guide to the Perplexed* and Thomas S. Kuhn's *Essential Tensions* are fundamentally in accord. And, to speak from personal experience, life is much harder for one who rejects their counsel – for one who is unable to accept as truly wise the person whom the community judges to be wise.

So much for background information. I apologize for my lengthy and round-about presentation, but only now do I feel comfortable – somewhat comfortable – coming to my point. Twenty odd years ago I wrote my *Towards an Historiography of science.* It was published in 1963 and reissued in 1967. It is still popular – perhaps my biggest public success to date. I was and still am very happy with its reception. Yet now, having to look back on it, I feel I should say what was found wrong with it and what I find wrong with its reception and what I do not like about its message.

My brief work presented the two paradigms concerning the writing of the history of the physical sciences, rejected them and proposed that the views of Karl Popper and some of the work of Alexandre Koyré offer a better paradigm. Most readers have concluded that I endorse Popper's paradigm with no qualification. Yet in that work I hint that I think I have a better paradigm, one which weds some ideas of my teacher Popper with ideas of Émile Meyerson and other aspects of the work of his already mentioned august disciple Alexandre Koyré. I have expouned this paradigm elsewhere – in my *Science in Flux,* for example.

Some historians of science dismissed my *Towards an Historiography of Science* as small fry, either because they followed a paradigm I was rejecting or because I was an outsider. Also some critics who said kind and appreciative words about it criticized it from the viewpoint of a given paradigm – for example Nicholas Rescher and Edwin G. Boring. I will not respond to this, since the quarrel between paradigms is an elaborate, wearisome matter. I should also mention that some of those who said kind and appreciative words about my book complained that I was an outsider, particularly Thomas S. Kuhn and Derek J. deSolla Price. And, I suppose, I am an outsider.

What characterizes an outsider, what is it to be an outsider, and why does it matter? These questions signify. I shall show this, discuss these questions, apply them to the place of my work in the literature and leave it for you to draw any conclusions you think appropriate.

That the question signifies I have already argued. Reforms, to repeat, are within a culture or a cultural community maintaining a certain social continuity. The social continuity, so goes the popular prejudice, is maintained by the society in question sharing a certain paradigm – sharing a certain prejudice, I will say. To remain both flexible and prejudiced the community needs an intelligent and a strong leadership. The leadership, then, and only the leadership, prescribes a paradigm-shift. Query: what if an ousider offers a paradigm shift? Without comparing myself to other outsiders, and for mere presentational purposes, it may be advisable to take the biggest examples around. Moses the law-giver is an example, as is Einstein the patent-tester. For more examples one can use Collin Wilson's *The Outsider,* though it is a cheap romantic essay, once extremely popular and now all but forgotten. Anyway, the usual way an outsider's views become incorporated is for them to be endorsed by the current leaders. Aaron endorsed Moses, Max Planck endorsed Albert Einstein. Smaller examples are more abundant but less useful as examples. Myself, I am no example at all. I was not endorsed by the leadership, and, when my dissatisfaction with my book is presented later on, my satisfaction with this fact will be apparent enough.

Why, then, the complaint that I am an outsider? If my views are useful and endorsed by the leadership or useless and rejected by the leadership, then there is no problem. Yet neither is the case. Why? I do not know, but I have a conjecture. The field of the history of science in general and of the history of the physical sciences in particular has no adequate leadership and no recognized leadership. True, there are the trappings of agreement about common and shared prejudices, trappings of a totem pole, of territorial claims respected by others, and so on. Yet all this is so unsatisfactory that it can be disregarded.

Unsatisfactory to the rank-and-file who need leaders, I mean; not to those who would rather get lost without a leader than follow one. It is an historical fact, easily ascertainable, that the field is still dominated by diverse paradigms, that a few authors are purists who stick consistently to the paradigm of their choice, and that a few write with disregard for all paradigms. Of course, those who follow one paradigm or no paradigm have the virtue of consistency, those who are eclectic may come up with exciting finds and thus invite others to clean their works of their inconsistency. And, on the whole, the Koyré-Popper paradigm I sided with is by now a reasonable contender in this multi-paradigm field, and I am proud of my little share in this development. But, not to lose my sense of proportion, let me observe that the

Kuhnian paradigm, not discussed in my *Historiography*, since it was presented to the world when that work was written, is more popular though not dominant: there is no dominant paradigm in the field.

Well, then, to my question, what characterizes an outsider, what makes one an outsider, and why does it matter? I think now the answer gets clear. In particular, the fact that I have published some books and research papers in the field of the history of science proper need not qualify me as an insider. Again I should mention a venerable precedent. In his later years Albert Einstein counted as a physicist for sure, yet being neither rank-and-file nor a leader he counted as an outsider. I still remember the sense of shock I experienced when I learned that my physics teacher, Julio Raccah, had no wish to read Einstein's latest papers. An outsider, then, is one who is neither leader or rank-and-file. Most outsiders, of course, simply play no roles in the field, but not all of them: those who have something to contribute not endorsed by the leadership may also be viewed as outsiders. What makes them outsiders is a different matter.

Of course, the interesting cases are the problematic ones. For example, a person may be a partial insider, as was Einstein. Michael Faraday's case was more complex: he was an insider experimentalist and an outsider theoretician. James Clerk Maxwell, his leading disciple, was surprised to learn that he was a theoretician. Yet this fact was kept secret, and historians of science had to rediscover it. Nevertheless, the fact that this fact was a secret is still not known. This is what is called, since Watergate, the cover-up of the cover-up. The cover-up includes the fact that reviews of my *Faraday as a Natural Philosopher* often failed to report this point, which is the book's message.

This brings me to the heart of the matter, namely to the significance of it all. Though I do not like intellectual leadership, I do recognize the existence of communities of scholars, and the need communities have for leadership. I also recognize the fact that communities are often identified by shared prejudices, by paradigms, but this fact is one I dislike and claim that it is no longer true in any pluralist society.

I therefore propose, first and foremost, to distinguish community from beliefs. Admittedly, communities often play certain roles and the leadership needs to know the consensus (on the roles) of their communities. Kuhn says the role of the rank-and-file physicist is to solve certain technological problems which serve the community at large: the theoretical investigations of the leading physicists, then, serve the rank-and-file in their discharge of their recognized task. If this is what Kuhn says, then he articulates a sentiment widely held by physicists in the period of large-scale United States Federal Government Research and Development Funding, roughly between the A-bomb and the big budgetary cuts. If so, Kuhn is by now out-of-date. If this is not what Kuhn says, then I do not know what he thinks the rank-and-file scientist's task of puzzle-solving is. In any case, this is an example of a question, what is the current role of the community of physicists? Perhaps also of the subsequent or the derivative question, what is the priority order of today's agenda within physics?

But I am talking of the history of physics, not of physics. And I have argued in my *Historiography* that the self-selected role of the historian of physics is to orchestrate the process of hero worship which the community at large is supposed to perform with the scientist as the hero. I find this task distasteful. I admire Popper and Koyré so because they have, following Einstein, taken it as a central idea that an error can be an excellent intellectual adventure and a great contribution to the progress of science. Yet this view is still quite unpopular, and the ascription of an error to a thinker is quite often taken as a way to belittle the contribution of that thinker as a matter of course. This is something I find intolerable and intolerably common in the literature of the history of science. This, I think, brands me an outsider in a clear manner: those who perform the task of presenting scientists as if they had never made an error do not like to be told what it is. Nor are they wrong: many social tasks have an ambiguity about them which is essential to their proper functioning.

This is not to say that such social tasks are enjoyable. They seldom are. And what I tried to show in my *Historiography* is that historians of science have better and more enjoyable tasks to perform, and some of them do. This put me in the position of an underground advisor for quite a few historians of science who, I know, have read my work with pleasure, and deem it beneficial for their work, yet on the condition that they pass over this fact in silence. As historians they know that future historians may, if they put their mind to it at all, discover some unmentioned facts. But as members of their own community they know what they could do and what not.

There is little doubt that a radical change took place in the writing of the history of the exact sciences in the last two decades. No doubt, Koyré and Popper are the chief sources of inspiration here. Even historians of science who reject Popper's ideas, such as Gerald Holton, admit this, not to mention Kuhn, whom I heard to say both that he has no quarrel with Popper and the contrary. Philosophers and historians of science, notably Dudley Shapere, have taken the accord on matters of principle between Popper and Kuhn despite surface difference, as the emergence of the new paradigm.

This is the time for me to quit. When I wrote my *Historiography* I defended Popper since he was an outsider. Now he is the insider and I want to disengage from him as sharply as I can. Here, briefly, is my central point of dissent written from the viewpoint of particular interest in the history of the exact sciences.

What use is there in raising a palm tree and taking an axe to it? What is gained by the game of conjectures and refutations which Popper declares science to be?

My major critique of Popper is that some conjectures are worthless though highly refutable and at times, indeed, refuted; some conjectures are technologically useful though scientifically worthless and they concern only engineers, or, at times, engineers and mathematicians, particularly ones concerned with complex differential equations, complex approximations and their theory, and more. But here I wish to stay a little with the useless, as the major point of criticism of Popper's views.

Popper opposes scholasticism, i.e., Talmudism, as do almost all philosophers of science. It is well-known that scholasticism dogmatically clings to a pertrified para-digm and merely adds to the paradigm patchwork upon patchwork. Dogmatism was characterized by Sir Francis Bacon as disregard for refutation and as making light of it and as meeting it by minimal adjustments. Amazingly, W. V. Quine endorsed this view. I say amazingly because the Duhem-Quine thesis is the claim that one can always both be dogmatic and accept criticism. The Duhem-Quine thesis, inciden-tally, is logically true and was proven by Duhem with ease.

If Popper advocates conjectures and refutations but rejects scholasticism, he should explain how. He does. He says, scientific conjectures are highly refutable, scholastic ones are not. This is a very beautiful claim; it is false. For example, it is possible to amend the Newtonian law of gravity by adding to the inverse square an inverse cube and other factors and indeed this is practiced by every space-program. The amendment is very useful, highly refutable, yet, scientifically viewed, it is quite scholastic. Einstein said repeatedly, he never considered it a scientific option. Popper, therefore, is in error.

I have argued elsewhere, that in science conjectures are chosen within a paradigm, until it gets tired and exhausted: it is no longer a source of inspiration for new conjectures that may explain older ones. The existence of hierarchies of theories as series of approximations is the reason we may consider them both as series of conjectures and refutations and as progress, both progress in explanation, and as progress in shifts from paradigm to better paradigm. The historians of science should not conceal refutation, but they may also ask, what ground was gained in our heroes having first raised a whole grove of palm trees and then having taken axes to them? That science progresses, that it progresses in diverse ways, is unquestionable at present. That this progress is one, but only one, factor making the history of science exciting, is, I think, quite obvious. My *Historiography* was more adequate as a guide. Yet if science is, among other things, a force liberating our thinking from past constraints, and if some individuals are helped by it to become autonomous, then the future autonomous historians of science will have enough exciting work on their hands.

Perhaps the point from my *Historiography* which I like best is that of asserting autonomy – of both the historians and their heroes, and by the clear contrast of the opinions they happen to represent. And, to my regret, till today too many historians – especially the large group of historians who are still under the pernicious influence of Henry Guerlac – identify with their heroes instead of sympathizing with them, worship them and beautify their results in violation of historical truth while pretend-ing to be scholars. We have to know that our heroes have viewed science quite differently from us, and present their work both from their viewpoints and ours. In writing my study of Faraday I attempted to do this, critically presenting his social background, his inner conflicts, his methodology and his paradigm – constantly reminding my reader that all of Faraday's discoveries are by now superseded. The ideal history of science today, I still think, is the very opposite of identifying one's ideas with the ideas of one's hero. We have to be problem-oriented and record that most past theories are refuted, and do so without being apologetic in the least. Yet this is not enough, and I wish to supplement my earlier work. We may want to contrast our own way of choosing problems with how past thinkers chose their problems and how they approached them. At times this would lead us to consider social history, at times the spirit of the age, including the popular prejudices shared by our heroes, including, particularly, their metaphysical and methodological views, and always in contrast with today's case. We may wish to offer integrated portraits – intellectual portraits – of our heroes, but we have to stay on guard: no person is fully integrated and portraits should not include more than history invites.

Some studies approach this idea. The current concerted study of Newton, in particular, comes close to it. But as yet too few historians notice that he was influenced by conflicting philosophies – by mystic doctrines, by inductive philosophy, by Cartesianism and more. To ascribe these views to him and ask how much they helped and how much they hindered his progress is still regrettably uncommon. The history of science is as open and challenging as ever.

# The Semantic Approach to Scientific Theories

BAS C. VAN FRAASSEN
*Princeton University*

The purpose of this paper is not to be new or original, but to provide a concise exposition of a certain approach in philosophy of science, at the hands of a loosely associated group of contemporary philosophers.\* I do not want anyone to suffer from guilt by asociation, so I should emphasize that any two authors mentioned here may agree on only a few aspects of the discussed approach, and disagree on others.

I locate the real beginning of the semantic approach, the place where it became consciously distinct, in E. W. Beth's *Natuurphilosophie* (1948), and summarized rather cryptically in two short articles in English and French in 1948/49. I read this book in 1965 and I gave an exposition of his views in my 'On the Extension of Beth's Semantics of Physical Theories' (1970). As I will explain below, some of the main ideas I found there were already in some way 'in the air' in North America. Since his view could be sloganized by saying that *the analysis of quantum logic provides the paradigm for the semantic analysis of physical theory,* it is clear that his view also had antecedents in the thirties and forties, flourishing perhaps entirely unconsciously of the implied opposition to the approach of the logical positivists. A good example is Herman Weyl's article 'The Ghost of Modality' (1940).

My own particular variant of the semantic approach is strongly colored by my attempts to become a more thoroughgoing empiricist, while such other advocates as Frederick Suppe and Ronald Giere are outspoken scientific realists. My exposition could not easily remain unbiased on this issue, but I hope that the essential neutrality thereon, of the semantic view, will be clear nevertheless.

## 1. The aim of science

Philosophy of science attempts to answer the question *What is science?* in just the sense in which philosophy of art, philosophy of religion, and the like answer the

---

\* The main content of this paper is accordingly a combination of several previous ones; mainly my 'Aim and Structure of Scientific Theories' and 'Theory Construction and Experiment'; see bibliography. A more complete defense of my own particular variant of the approach will be found in *Images of Science*, edited by P. M. Churchland and C. A. Hooker, Chicago: University of Chicago Press, 1985.

*Nancy J. Nersessian (editor), The Process of Science. ISBN 90-247-3425-8.*
© *1987, Martinus Nijhoff Publishers (Kluwer Academic Publishers), Dordrecht. Printed in the Netherlands.*

similar question about their subject. For better or for worse our tradition has focussed on the scientific theory rather than on scientific activity itself (on the product, rather than on the aim, conditions, and process of production, to draw an analogy, which is already one that points in its terminology to the product as most salient feature). Yet all aspects of scientific activity must be illumined if the whole is to become intelligible. I shall therefore devote a preliminary section to the aim of science, and to the proper form of epistemic or doxastic attitudes toward scientific theories, before entering upon a description of their structure.

The activity of constructing, testing, and refining of scientific theories – that is, the production of theories to be accepted within the scientific community and offered to the public – what is the aim of this activity?

I do not refer here either to the motives of individual scientists for participating, or the motives of the body civic for granting funds and otherwise supporting the activity. Nor do I ask for some theoretically postulated 'fundamental project' which would explain this activity. It is part of the straightforward description of any activity, communal or individual, large-scale or small, to describe the end that is pursued as one of its defining conditions. In the most general terms, the end pursued is success, and the question is what counts as success, what are the criteria of success in this particular case?

We cannot answer our particular questions here without some reflection on what sort of thing this product, the scientific theory, is. A scientific theory must be the sort of thing that we can accept or reject, and believe or disbelieve; accepting a theory implies the opinion that it is successful; science aims to give us acceptable theories. To put it more generally, a theory is an object for epistemic or at least doxastic attitudes. A typical object for such attitudes is a proposition, or a set of propositions, or more generally a body of putative information about what the world is like, what the facts are. If anyone wishes to be an instrumentalist, he has to deny the appearances which I have just described. An instrumentalist would have to say that the apparent expression of a doxastic attitude toward a theory is elliptical; 'to believe theory T' he would have to construe as 'to believe that theory T has certain qualities'. I shall not follow that path. Let me state here at once, as a first assumption, that the theory itself is what is believed or disbelieved.

At this point we readily can see that there is a very simple possible answer to all our questions, the answer we call *scientific realism*. This philosophy says that a theory is the sort of thing which is either true or false; and that the criterion of success is truth. As corollaries we have that acceptance of a theory as successful is, or involves the belief that it is true; and that the aim of science is to give us (literally) true theories about what the world is like.

That answer must of course be qualified in various ways to allow for our epistemic finitude and the consequent tentativeness of reasonable doxastic attitudes. Thus we add that although it cannot generally be *known* whether or not the criterion of success has been met, we may reasonably have a high degree of belief that it has been, or that it is met approximately (i.e., met exactly by one member of a set of 'small variants' of the theory), and this imparts similar qualifications to acceptance in practice. And we add furthermore of course that empiricism precludes dogmatism, that is, *whatever* doxastic attitude we adopt, we stand ready to revise in the face of further evidence. These are all qualifications of a sort that anyone must acknowledge, and should therefore really go without saying. They do not detract from the appealing and as it were pristine clarity of the scientific realist position.

I did not want to discuss the structure of theories before bringing this position into the open, and confronting it with alternatives. For it is very important, to my mind, to see that an analysis of theories – even one that is quite traditional with respect to what theories are – does not presuppose it. Let us keep assuming with the scientific realist, that theories are the sort of thing which can be true or false, that they say what the world is like. What they say may be true or false, but it is nevertheless literally meaningful information, in the neutral sense in which the truth value is 'bracketed'.

There are a number of reasons why I advocate an alternative to scientific realism. One point is that reasons for acceptance include many which, *ceteris paribus,* detract from the likelihood of truth. In constructing and evaluating theories, we follow our desires for information as well as our desire for truth. For belief, however, all but the desire for truth must be 'ulterior motives'. Since therefore there are reasons for acceptance which are not reasons for belief, I conclude that acceptance is not belief. It is to me an elementary logical point that a more informative theory can not be more likely to be true – and attempts to describe inductive or evidential support through features that require information (such as 'inference to the best explanation') must either contradict themselves or equivocate.

It is still a long way from this point to a concrete alternative to scientific realism. Once we have driven the wedge between acceptance and belief, however, we can reconsider possible ways to make sense of science. Let me just end these preliminary remarks now by stating my own position, which I call *constructive empiricism.* It says that the aim of science is not truth as such but only *empirical adequacy,* that is, truth with respect to the observable phenomena. Acceptance of a theory involves as belief only that the theory is empirically adequate – but it involves more than belief.

While truth as such is therefore, according to me, irrelevant to success for theories, it is still a category that applies to scientific theories. Indeed, the *content* of a theory is what it says the world is like; and this is either true or false. The applicability of this notion of truth value remains here, as everywhere, the basis of all logical analysis. When we come to a specific theory, there is an immediate philosophical question, which concerns the content alone: *how could the world possibly be the way this theory says it is?*

This is for me the foundational question *par excellence.* And it is a question whose discussion presupposes no adherence to scientific realism, nor a choice between its alternatives. This is the area in philosophy of science where realists and anti-realists can meet and speak with perfect neutrality.

## 2. Theory structure – models and their logical space

In this section I shall present a view of theories which makes language largely irrelevant to the subject. Of course, to present a theory, we must present it in and by language. That is a trivial point, for any effective communication proceeds by language, except in those rare cases in which information can be conveyed by the immediate display of an object or happening. In addition, both because of our own history – the history of philosophy of science which became intensely language oriented during the first half of this century – and because of its intrinsic importance, we cannot ignore the language of science. Hence I shall make it the subject of the last section.

In what is now called the received view, a theory was conceived of as an axiomatic theory. That means, as a set of sentences, defined as the class of logical consequences of a smaller set, the axioms of that theory. A distinction was drawn: since the class of axioms was normally taken to be effectively presentable, and hence syntactically describable, the theory could be thought of as in itself uninterpreted. The distinction is then that scientific theories have an associated interpretation, which links their terms with their intended domain. We all know the story of misery and pitfalls that followed. Only two varieties of this view of scientific theories as a special sort of interpreted theories, emerged as anywhere near tenable. The first variety insists on the formal character of the theory as such, and links it to the world by a partial interpretation. Of this variety the most appealing to me is still Reichenbach's, which said that the theoretical relations have *physical correlates*. Their partial characters stand out when we look at the paradigm example: light rays provide the physical correlate for straight lines. It will be immediately clear that not every line is the path of an actual light ray, so the language-world link is partial. The second variety, which came to maturity in Hempel's later writings, hinges for its success on treating the axioms as already stated in natural language. The interpretative principles have evolved into axioms among axioms. This means that the class of axioms may be divided into those which are purely theoretical, in which all non-logical terms are ones specially introduced to write the theory, and those which are mixed, in which non-theoretical terms also appear. It will be readily appreciated that in both these developments, despite lip service to the contrary, the so-called problem of interpretation was left behind. We do not have the option of interpreting theoretical terms – we only have the choice of regarding them as either (a) terms we do not fully understand but know how to use in our reasoning, without detriment to the success of science, or (b) terms which are now part of natural language, and no less well understood than its other parts. The choice, the correct view about the meaning and understanding of newly introduced terms, makes no practical difference to philosophy of science, as far as one can tell. It is a good problem to pose to philosophers of language, and to leave them to it.

In any tragedy,[1] we suspect that some crucial mistake was made at the very beginning. The mistake, I think, was to confuse a theory with the formulation of a theory in a particular language. The first to turn the tide was Patrick Suppes, with his well-known slogan: the correct tool for philosophy of science is mathematics, *not* metamathematics. This happened in the fifties – bewitched by the wonders of logic and the theory of meaning, few wanted to listen. Suppes' idea was simple: *to present a theory, we define the class of its models directly*, without paying any attention to questions of axiomatizability, in any special language, however relevant or simple or logically interesting that might be. And if the theory as such, is to be identified with anything at all – if theories are to be reified – then a theory should be identified with its class of models.[2]

This procedure is in any case common in modern mathematics, where Suppes had found his inspiration. In a modern presentation of geometry we find not the axioms of Euclidian geometry, but the definition of a Euclidean space. Similarly Suppes and his collaborators sought to reformulate the foundations of Newtonian mechanics, by replacing Newton's axioms with the definition of a Newtonian mechanical system. This gives us, by example, a *format* for scientific theories. In Ronald Giere's recent encapsulation, a theory consists of (a) the *theoretical definition,* which defines a certain class of systems; (b) a *theoretical hypothesis,* which asserts that certain (sorts of) real systems belong to that class.

This is a step forward in the direction of less shallow analysis of the structure of a scientific theory. The first level of analysis addresses the notion of theory *überhaupt,* but we do not want to stop there. We can go still a bit further by making a division between relativistic and non-relativistic theories. In the latter, the systems are physical entities developing in time. They have accordingly a space of possible states, which they take on and change during this development. This introduces the idea of a cluster of models united by a common *state-space;* each has in addition a domain of objects plus a 'history function' which assigns to each object a history, i.e., a trajectory in that space. A real theory will have many such clusters of models, each with its state-space. So the presentation of the theory must proceed by describing a class of *state-space types.*

In the case of relativistic theories, early formulations can be described roughly as relativistically invariant descriptions of objects developing in time – say in their proper time, or in the universal time of a special cosmological model (e.g., Robertson-Noonan models). A more general approach, developed by Glymour and Michael Friedman, takes space-times themselves as the systems. Presentation of a space-time theory T may then proceed as follows: a *(T-)space-time* is a four dimensional differentiable manifold M, with certain geometrical objects (defined on M) required to satisfy the *field equations* (of T), and a special class of curves (the possible trajectories of a certain class of physical particles) singled out by the *equations of motion* (of T).

Clearly we can further differentiate both sorts of theories in other general ways, for example with respect to the stochastic or deterministic character of imposed laws. (It must be noted however that except in such special cases as the flat space-time of special relativity – its curvature independent of the matter-energy distribution – there are serious conceptual obstacles to the introduction of indeterminism into the space-time picture.)

I must leave aside for now the details of foundational research in the sciences. But I want to point out that the point of view which I have been outlining – the *semantic view* as opposed to the received view – is much closer to practice there. The scientific literature on a theory makes it relatively easy to identify and isolate classes of structures to be included in the class of theoretical models. It is on the contrary usually quite hard to find laws which could be used as axioms for the theory as a whole. Apparent laws which frequently appear are often partial descriptions of special subclasses of models, their generalization being left vague and often shading off into logical vacuity. Let me give two examples. The first is from quantum mechanics: *Schrödinger's equation*. This is perhaps its best known and most pervasively employed law – but it cannot very well be an axiom of the theory since it holds only for conservative systems. If we look into the general case, we find that we can prove the equation to hold, for some constant Hamiltonian, under certain conditions – but this is a metamathematical fact, hence empirically vacuous. The second is the Hardy-Weinberg law in population genetics. Again, it appears in any foundational discussion of the subject. But it could hardly be an axiom of the theory, since it holds only under certain special conditions. If we look into the general case, we find a logical fact: that certain assumptions imply that it describes an equilibrium which can be reached in a single generation, and maintained. The assumptions are very special, and more complex variants of the law can be deduced for more realistic assumptions – in an open and indefinite sequence of sophistications.

What we have found, in this approach, is a way to describe relevant structures in ways that are also directly relevant, and seen to be relevant, to our subject matter. The scholastically logistical distinctions that the logical positivist tradition produced – observational and theoretical vocabulary, Craig reductions, Ramsey sentences, first-order axiomatizable theories, and also projectible predicates, reduction sentences, disposition terms, and all the unholy rest of it – had moved us *mille milles de toute habitation scientifique,* isolated in our own abstract dreams. Since Suppes' call to return to a non-linguistic orientation, now about thirty years ago, we have slowly regained contact.

### 3. Theory structure – relation to the world

Above I mentioned Giere's elegant capsule formulation of the semantic view: a theory is presented by giving the definition of a certain kind (or kinds) of systems plus one or more hypotheses to the effect that certain real (kinds of) systems belong to the defined class(es). We speak then of the *theoretical definition* and the *theoreti-*

*cal hypothesis* which together constitute the given formulation (in, so to say, canonical form) of the theory. A 'little' theory might for example define the class of Newtonian mechanical systems and assert that our solar system belongs to this class.

Truth and falsity offer no *special* perplexities in this context. The theory is true if those real systems in the world really do belong to the indicated defined classes. From a logical, or more generally semantic point of view, we may consider as implicitly given models of the world as a whole, which are as the theoretical hypotheses say it is. There is of course a very large class of models of the world as a whole, in which our solar system is a Newtonian mechanical system. In one such model, nothing except this solar system exists at all; in another the fixed stars also exist, and in a third, the solar system exists and dolphins are its only rational inhabitants. Now the world must be one way or another; so the theory is true if the real world itself is ( or is isomorphic to) one of these models. This is equivalent to either of two familiar sorts of formulations of the same point: the theory is true exactly if (a) one of the possible worlds allowed by the theory is the real world; or (b) all real things are the way the theory says they are.

But while the subject of truth yields no special conceptual difficulties in this context, I do not believe that it marks the relation to the world, which science pursues in its theories. This, as you will recall, is the point at issue between scientific realism and empiricism. Leaving the issue itself aside, I think that even scientific realists need to be acutely interested in a much closer, more empirical relation of theory to world. I call this relationship *empirical adequacy*.

The logical positivist tradition gave us a formulation of such a concept which was not only woefully inadequate but created a whole cluster of 'artifact problems' (by this I mean, problems which are artifacts of the philosophical approach, and not inherent in its subject). In rough terms, the empirical content of a theory was identified with a set of sentences, the consequences of that theory in a certain 'observational' vocabulary. In my own studies, I first came across formulations of more adequate concepts in the work of certain Polish writers (Przeleweski 1969, Wojcicki 1974), of Dalla Chiara and Toraldo di Francia (1973 and 1977) and finally of course in Patrick Suppes' own writings on what he calls empirical algebras and data models (1967, 1969). While some of these formulations were still more language-oriented than I liked, the similarity in their approach was clear: certain parts of the models were to be identified as *empirical substructures,* and these were the candidates for representation of the observable phenomena which science can confront within our experience.

At this point I perceived that the relationship thus explicated corresponds exactly to the one Reichenbach attempted to identify through this concept of coordinative definitions, once we abstract from the linguistic element. Thus in a space-time the geodesics are the candidates for the paths of light rays and particles in free fall. More generally, the identified spatio-temporal relations provide candidates for the relational structures constituted by actual genidentity and signal connections. These actual physical structures are to be embeddable in certain substructures of space-time, which allows however for many different possibilities, of which the actual is, so to say, some arbitrary fragment.

Thus we see that the empirical structures in the world are the parts which are at once *actual* and *observable;* and empirical adequacy consists in the embeddability of all these parts in some single model of the world allowed by the theory.

Patrick Suppes has very carefully investigated the construction of data models, and the empirical constraint they place on theoretical models. Thought of as concerned with exactly this topic, much apparently 'a prioristic' theorizing on the foundations of physics takes on a new intelligibility. A reflection on the possible forms of structures definable from joint experimental outcomes yields constraints on the general form of the models of the theories 'from below' which can then be narrowed down by the imposition of postulated general laws, symmetry constraints, and the like, 'from above'.

### 4. Theorizing: data models and theoretical models

New theories are constructed under the pressure of new phenomena, whether actually encountered or imagined. By 'new' I mean here that there is no room for these phenomena in the models provided by the accepted theory. There is no room for a mutable quantity with a discrete set of possible values in the models of a theory which says that all change is continuous. In such a case the old theory does not allow for the phenomenon's description, let alone its prediction.

I take it also that the response to such pressure has two stages, logically if not chronologically distinguishable. First the existing theoretical framework is widened so as to allow the possibility of those newly envisaged phenomena. And then it is narrowed again, to exclude a large class of the thereby admitted possibilities. The first move is meant to ensure empirical adequacy, to provide room for all actual phenomena, the rock-bottom necessary condition of success. The second move is meant to regain empirical import, informativeness, predictive power.

It need hardly be added that the moves are not made under logical compulsion. When a new phenomenon, say X, is described it is no doubt possible to react with the assertion that if it looked as if X occurred, one would only conclude that familiar fact or event Y had occurred. A discrete quantity can be approximated by a continuous one, and an underlying continuous change can be postulated. From a purely logical point of view, it will always be up to the scientists to take a newly described phenomenon seriously or to dismiss it. Logic knows no bounds to *ad hoc* postulation. This also brings out the fact of creativity in the process that brings us the phenomena to be saved. Ian Hacking put this to me in graphic terms when he described the quark hunters as seeking to create new phenomena. It also makes the point long emphasized by Patrick Suppes that theory is not confronted with raw data but with models of the data, and that the construction of these data models is a sophisticated and creative process. To these models of data, the dress in which the debutante phenomena make their debut, I shall return shortly.

In any case, the process of new theory construction starts when described (actual or imagined) phenomena are taken seriously as described. At that point there certainly is logical compulsion, dimly felt and, usually much later, demonstrated. Today Bell's Inequality argument makes the point that certain quantum mechanical phenomena cannot be accommodated by theories which begin with certain classical assumptions. This vindicates, a half century after the fact, the physicists' intuition that a radical departure was needed in physical theory.

Of the two aspects of theorizing, the widening of the theoretical framework, and it's narrowing to restore predictive power, I wish here to discuss the former only. There we see first of all a procedure so general and common that we recognize it readily as a primary problem-solving method in the mathematical and social as well as the natural sciences, any place where theories are constructed, including such diverse areas familiar to philosophers as logic and semantics. This method may be described in two ways: as *introducing hidden structure,* or 'dually' as *embedding.* Here is one example.

Cartesian mechanics hoped to restrict its basic quantities to ones definable from the notions of space and time alone, the so-called kinematic quantities. Success of the mechanics required that later values of the basic quantities depend functionally on the earlier values. There exists no such function. Functionality in the picture of nature was regained by Newton, who introduced the additional quantities of mass and force. Behold the introduction of hidden parameters.

The world 'hidden' in 'hidden parameters' does not refer to lack of experimental access. It signifies that we see parameters in the solution which do not appear in the statement of the problem.

We can 'dually' describe the solution as follows: the kinematic relational structures are embedded in structures which are much larger – larger in the sense that there are additional parameters (whether relations or quantities or entities). *The phenomena are small but chaotic; they are treated as fragments of a 'whole' that is much larger but orderly and simple.* This point could, I believe, be illustrated by examples from every stage of the history of science. When a point has such generality one assumes that it must be banal, and carry little insight. In such a general inquiry as ours, however, perspective is all; and we need general clues to find a general perspective. (That this particular point may be productive of more specific insights is in any case not an unreasonable hope. The most spectacular recent theoretical development may well be the deduction of Maxwell electrodynamics, Einstein geometrodynamics, and the Yang-Mills quark-binding field dynamics from the requirement of embeddability in space-time, by Hojman, Kuckar and Teitelboim.)

## 5. Theorizing illustrated

In order to illustrate the general view of what theorizing is like, which I have just presented, and to try and persuade you that it is a reasonable one, I shall now describe some recent activities in the foundations of quantum mechanics.

The whole point of having theoretical models is that they should fit the phenomena, that is, fit the models of data. So we need to look at what the latter must be like in general. Hence the development of Randall and Foulis', 'empirical logic'; Mackey, Jauch, and Piron's preliminary discussions of experimental questions; Ludwig and Mielnik's filters; and so forth. These authors write sometimes as if their program is one of transcendental deduction: study what the data models must be like, deduce what structure theoretical models must have if the data models are to be deducible, demonstrate the basic axioms of quantum mechanics as corollaries to this deduction. Since the theory has clear empirical content, success can be, at most, partial; but it is astonishing how much can be achieved in this way. Moreover the very fact that success must necessarily be partial is what gives the approach its value for the future: it brings within our ken alternatives to the extant theory that can rival it.

In Suppes' description, the experimentalist brings to the theoretician a small relational structure, constructed carefully from selected data. The examples Suppes mentions are specific and the little structures are algebras; hence he calls them 'empirical algebras'. The authors in quantum mechanics point to such small structures that represent data, and they are not always algebras; they are most generally partial algebras, or just partially ordered sets with some operations. Let us see how this happens.

In the typical sort of experiment discussed in connection with the Einstein-Podolski-Rosen paradox and Bell's Inequalities, we have two apparatus, L (for *'left')* and R. Each has (say) three settings or orientations; let, for instance, L1 be the proposition that L has been given the first setting. The experiments have each (say) two distinct possible outcomes, which we may represent by the numbers zero and one. Let, for instance, L30 be the proposition that L has the third setting and outcome zero.

When we carry out a particular run on this dual apparatus, we can give a score of T (for 'true') or F to some of these propositions. For example, the first time we do it, each apparatus was placed in the first setting; L had outcome 1 and R had outcome 0. An experimental report looks, in part, as follows:

| Proposition | Score |
|---|---|
| L1 | T |
| L2 | F |
| R1 | T |
| L10 | F |
| L20 | no score |
| R10 | T |

We note that L20 received no score. It could have been give F, simply on the basis that L had not been given the second setting. But this is useless information, and does not appear in the experimental report.

This single report is not likely to come to the theoretician's desk. What reaches him rather are reports of the form:

A) With initial preparation X, the probability of outcome Lia, given setting Li, equals r.

B) For all initial preparations, the probability of (Lia & Ria), given settings Li and Ri, equals zero.

There was an extrapolation before these conclusions were reached: the extrapolation from found relative frequencies to probabilities. This is regarded as no different from the extrapolation of data points on a graph, to a smoothed curve.

But the report that comes in forms (A) and (B) leads us to a mathematical structure that may properly be called the *data model*. The important relation stated between Lia and Ria in (B) is that when they can receive an informative score at all (i.e., when the preconditions Li and Ri obtain), they cannot both receive the score T. We then call Lia and Ria *orthogonal*. It also means that Ri1 must receive score T when Li0 does (modulo probability zero), and again when the informative scoring conditions obtain; and we call that *implication*. The latter is a partial ordering, and so we have here a partially ordered set *(poset)* with an orthogonality relation.

Reflection on this form of representation leads to assertions of the form: all data models can take the form ... A popular way to fill in the dots is to say 'ortho-poset' (i.e., poset with orthocomplement). A. R. Marlow, whose work I am about to take as a special example, used the more general characterization *dual poset;* that is, a partially ordered set with zero element and equipped with a single operation, *duality* $(x \neq x'; x = x''; $ and $x$ implies y only if y' *implies* x').

In the world of mathematical entities there are many dual posets. Widening our theoretical framework will consist in the provision of models that can have very strange dual posets of experimental propositions embedded. But the embedding must be good, that is, we must be able to see in the theoretical model all the significant features. I mentioned parts of the experimental report labelled (A). They start 'With initial preparation X, . . .' and then they mention probabilities. These probabilities characterize what is called the *state* prepared by procedure X. And these states (they look like fragments of ordinary probability functions, in that they assign probabilities only to propositions for which the informative scoring conditions obtain) must be 'visible' in a certain sense in the computational structure of the theoretical model.

Here is Marlow's theorem. It requires two preliminary definitions. A *probability function* on a dual poset is any function f with the properties $f(0) = 0$, $f(x') = 1 - f(x)$, and $f(x) \leqslant f(y)$ if x implies y. A *base* for the dual poset is any set B of elements which does not contain the zero element, nor does it contain two orthogonal elements (here defined by the relation (x *implies* y$'$)), but does contain either x or x$'$ for each x. (Remark: every set with the first two properties can be extended to one that has all three. Note also that any set of elements that have all received score T on a particular occasion is intuitively required to have the first two properties.) The theorem says now that if we have a dual poset and a base, we can embed the poset in the algebra of projection operators on a Hilbert space, in such a way that duality becomes ortho-complementation, the partial ordering is preserved for elements within the base, and each probability function on the poset can be associated with a vector and becomes calculable by means of the familiar trace computation used in quantum mechanics.

This is an extraordinarily general result. Marlow takes the result as justification for his project to write space-time theories in Hilbert space formalism. The result provides good reason, after all, to write all physical theory in that mathematical framework. Of course he realizes that in some ways the theorem is less than totally general (the implication order is preserved only within the base!) and in some ways less than informative (there are enormous Hilbert spaces with room to embed almost anything) but the postulates he adds, and intends to add, will narrow down the *embarras de richesse* to recover empirical content.

Let us look, however, as second main illustration, at a line of thought born from dissatisfaction with the way phenomenal structures are embeddable in the Hilbert space framework. I refer to *operational quantum mechanics,* associated with Ludwig, Mielnik, Davies, and Edwards.

To explain how data models can take the form of ortho-posets, or more generally dual posets, I already gave a brief sketch of a quite typical experimental set-up, of an intermediate degree of complexity. Let me now start an alternative sketch; the two will not be incompatible. In a typical simple test, a system is *prepared* in a certain way; an *operation* is performed on it; and a *question* is asked. In the simplest case the question has *yes* or *no* as possible answers. We may keep count, as we repeat the text, of how often we receive *yes* as answer. We visualize the situation by imagining a *source* which sends out a beam of particles, that encounter a *barrier,* and a *counter* on the other side of the barrier that clicks every time a particle reaches it. This is a good picture for it has all the general features indicated; we appear to lose no generality if we focus attention on it.

The barrier affects the intensity of the beam; for example, if the barrier were not there, the counter would be clicking twice as rapidly. The sort of barrier determines the sort of question being asked. When the situation, so conceived, is embedded in the mathematical apparatus of Hilbert space, each source has an associated statistical operator W (representing the *prepared state)*, and each barrier an associated projection operator P (representing the *question asked);* the probability of a single particle passing the barrier (the factor by which the beam intensity is diminished) is calculated by the Born rule as $p = Tr(PW)$.

Many situations of the general sort described can indeed be modelled in this way. But others have to be treated in more round-about fashion. Intuitively we say that there is an *observable* in such a set-up. Whether that is so in the sense now instilled in us by quantum theory, has a simple criterion: there must be a state such that the *yes* answer becomes certain, that is, receives probability one. That state is then called an eigenstate of the observable being measured.

It is certainly possible to find examples where it looks intuitively as if we are measuring something, but the criterion is not met. Suppose we place an atom of a radioactive substance near a Geiger counter, and ask *Will the counter click within four to five minutes from now?* Is that not a simple form of measurement, of something we could call, say, the decay time of the atom? Yet we cannot prepare the atom in a state so as to make the *yes* answer certain.

A clearer example, first introduced into the literature, I think, by Shimony, and discussed especially by the Dutch authors de Muynk, Cooke, and Hilgevoort, is pictured in Figure 1. We have a battery of three Stern-Gerlach apparatus, testing for spin. They are so arranged that particles exiting from the first, along the top channel, encounter the second, while those exiting in the first top channel will exit along the second top channel, and so forth. We may choose the orientations so that those transition probabilities equal 1/2. As is easily seen in the diagram, there is no initial state which makes certain an exit along the second top channel; for the probability of

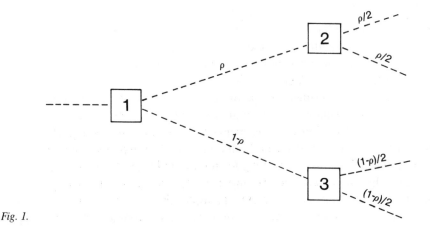

*Fig. 1.*

that exit equal p/2, which has maximum value 1/2. Here we have a question for which the *yes* answer *cannot* be certain. We can at most represent it by means of a sequence of questions, of the directly representable sort. Perhaps we could say that a 'derived observable' has been measured.

Operational quantum mechanics may be thought of as enlarging the theoretical models so as to allow the embedding of all such empirical structures in the same way. The questions asked by our battery of apparatus are treated on a par with those asked by a single apparatus. This enlargement proceeds as follows. We represent the (ensemble prepared by the) source by means of two parameters: the statistical operator W and a number $t$ which represents the production intensity. The barrier then affects both state W and intensity $t$, changing W to some other state W', and reducing the intensity $t$ to $rt$ (with $0 \leq r \leq 1$). We may set the initial intensity equal to 1, so that $t$ itself must belong to the interval $(0, 1)$. Referring to the statistical operators W – positive definite Hermitean operators with unit trace – as the 'old states', we represent the new states by their multiples $r$W. These are not in general themselves statistical operators, for if W has unit trace then $r$W has trace $r$.

Mathematically the enlargement proceeds in several steps. We begin with the 'ball' S of old states; generate the real cone

$$B+ = \{rW : W \varepsilon S \text{ and } 0 \leq r \leq 1\}$$

and then enlarge that cone by closing it under the usual linear operations, thus forming a real vector space B, a Banach space with the trace as norm. The physical operations can now be represented uniformly, by considering their effects on 'old state' and trace/intensity.

These two longish examples I take to reveal philosophically especially significant aspects of scientific theorizing. I turn to experimentation.

## 6. Experimentation: as test and as means of inquiry

Theory construction I have described as being ideally divisible into two stages: the construction of sufficiently rich models to allow for the possibility of described phenomena, and the narrowing down of the family of models so as to give the theory greater empirical content. There must be a constant interplay between the theoretician's desk and the experimenter's laboratory. Here too I wish to distinguish two aspects, of which I shall discuss one at greater length than the other.

The well-known function of experimentation is hypothesis testing. The experimenter reads over the theoretician's shoulder, and designs experiments to test whether the narrowing down has not gone too far and made the theory empirically inadequate. This characterization is simple and appealing; it is unfortunately over-simple. It overlooks first of all the fact that hypothesis testing is in general comparative, and ends not with support or refutation of a single hypothesis but with support of one hypothesis against another. It overlooks secondly that testability has to do with informativeness; a theory may be empirically adequate – or at least adequate with respect to a certain class of phenomena – but not sufficiently informative about them to allow the design of a test which could bring it support. Hence when we try to evaluate theories on the basis of their record even in the most assiduous testing, the ranking will reflect not only adequacy or truth, but also informativeness. Certainly these are both virtues, and both epistemic values to be pursued (as has been cogently argued by Isaac Levi), but it means that it is a mistake to think in terms of pure confirmation. To give an analogue: the number of votes a candidate receives is a measure of voter support for his platform but it is also a function of media exposure, and so it is not a pure measure of voter support. These reflections clearly bear on Glymour's theory of testing and relevant evidence, and his use of this important and original theory in arguments concerning scientific realism; but I have pursued them elsewhere.

There is another function of experimentation which is less often discussed and in the present context more interesting.

This second function is the one we describe in the language of discovery. Chadwick discovered the neutron, Millikan the charge of the electron, and Livingstone the Zambesi river. Millikan's case is a good illustration. He observed oil droplets drifting down in the air between two plates, which he could connect and disconnect with a battery. By friction with the air, the droplets could acquire an electrostatic charge; and Millikan observed their drifting behavior, calculating their apparent charges from their motion. Thus he found the largest number of which all apparent charges were integral multiples, and concluded that number to be the charge of a single electron. That number was *discovered*. The number of such charges per Faraday equals Avogadro's number! No one could have predicted that! You would think that empiricists would be especially bothered by such a scientific press release. For it appears that by carefully designed experiment we can discover facts about the unobservable entities behind the phenomena.

We must note first however that the division of experiments into means of testing and means of discovery is a division neither by experimental procedure nor by experimental apparatus. The set-up and operations performed would have been just the same if Millikan had made bold conjectures beforehand about that number, and had set out to test these conjectures. The division is rather by function *vis à vis* the ongoing process of theory construction.

The theory is written, so to say, step by step. At some point, the principles laid down so far imply that the electron has a negative charge. A blank is left for the magnitude of the charge. If we wish to continue now, we can go two ways. We could certainly proceed by trial and error, hypothesizing a value, testing it, offering a second guess, testing again. But alternatively, we can let the experimental apparatus write a number in the blank. What I mean is: in this case the experiment shows that unless a certain number (or a number not outside a certain interval) is written in the blank, the theory will become empirically inadequate. For the experiment has shown by actual example, that no other number will do; that is the sense in which it has filled in the blank. So regarded, *experimentation is the continuation of theory construction by other means.*

Recalling the similar saying about war and diplomacy, I should like to call this view the 'Clausewitz doctrine of experimentation'. It makes the language of construction, rather than of discovery, appropriate for experimentation as much as for theorizing.

## 7. The language(s) of science

We arrive now finally at the subject which I have mostly tried to banish from our discussion: language. I must admit that I too was, to begin, overly impressed by certain successes of modern logic. Thus one reviewer (John Worrall) of my book *The Scientific Image,* was able to quote the remark in my first paper on the subject, that the interrelations between the syntactic and semantic characterizations of a theory 'make implausible any claim of philosophical superiority for either approach'. The interrelations referred to are of course those described by the generalized completeness proof. I have long since changed my mind about its significance, both theoretical and practical.

To begin with the theoretical point, when a theory is presented by defining the class of its models, that class of structures cannot generally be identified with an elementary class of models of any first-order language.[3] The reason is found in the limitative meta-theorems, which brought to light the dark side of completeness. To take only the most elementary example, if a scientist describes a class of models, the mathematical object he is most likely to include is the real number continuum. There is no elementary class of models of a denumerable first-order language each of which includes the real numbers. As soon as we go from mathematics to metamathematics, we reach a level of formalization where many mathematical distinctions cannot be captured – except of course by *fiat,* as when we speak of 'standard' or 'intended' models. The moment we do so, we are using a method of description not accessible to the syntactic mode.

On the practical side we must mention the enormous distance between actual research on the foundation of science, and syntactically capturable axiomatics. While this disparity will not affect philosophical points which hinge only on what is possible 'in principle', it may certainly affect the real possibility of understanding and clarification.

Given this initial appreciation of the situation, shall we address ourselves to language at all? The answer I think is *yes,* not on any general grounds, but for a number of specific reasons.

Before detailing those specific reasons, let us look for a moment at language and the study of language in a general way. Russell made familiar to us the idea of an underlying *ideal language*. This is the skeleton, natural language being the complete living organic body built on this skeleton, the flesh being of course rather accidental, idiosyncratic, and molded by the local ecology. The skeleton, finally, is the language of logic; and for Russell's contemporaries the question was only whether *Principia Mathematica* needed to be augmented with some extra symbols to fully describe the skeleton.

Against this we must advance the conception of natural language as not being constituted by any one realization of any such logical skeleton. Logic has now provided us with a great many skeletons. Linguists have uncovered fragments of language in use for which no constructed logical skeleton yet provides any satisfactory model. Natural language consists in the resources we have for playing many different possible language games. Languages studied in logic texts are models, rather shallow models, of some of these specific language games, some of these fragments. To think that there must in principle exist a language in the sense of the objects described by logic, which is an adequate model for natural language taken as a whole, may be strictly analogous to the idea that there must exist a set which is the universe of set theory.

So if we now apply our logical methods in the philosophy of science we should, as elsewhere, set ourselves the task of modelling interesting fragments of language specially relevant to scientific discourse. These fragments may be large or small.

In my opinion, choosing the task of describing a language in which a given theory can be formulated, is a poor choice. The reason is that descriptions of structure in terms of satisfaction of sentences is, as far as I can see, generally less informative and less illuminating than direct mathematical (instead of meta-mathematical) description. It is the choice, explicit or implicit, to be formed in almost all linguistically oriented philosophical studies of science. It was the implicit choice behind, certainly, almost all logical positivist philosophy of science.

At the other extreme, we may choose a very small fragment, such as what I have called the fragment of *elementary statements*. Originally I characterized these as statements which attribute some value to a measurable physical magnitude. The syntactic form was therefore trivial – it is always something like 'm has value r' – and therefore the semantic study alone has some significance. Under pressure of various problems in the foundations of quantum mechanics, I broadened my conception of elementary statements in two ways. First, I admitted as possibly logically distinct the attributions of ranges (or Borel sets) of values. Second, I admitted as possibly distinct the attribution of states of certain sorts, on the one hand, from that of values to measurable magnitudes. (It should be added however that I soon found it much more advantageous to concentrate on the propositions expressible by elementary statements, rather than on the statements themselves. At that point there is not even a bow in the direction of syntactic description.)

There are points between these two extremes. I would point here especially to certain forms of natural discourse that are prevalent in the informal presentation of scientific theory, but which have a long history of philosophical perplexities. The main examples are causality and physical modality. From an empirical point of view, there are besides relations among actual matters of fact, only relations among words and ideas. Yet causal and modal locutions appear to introduce relations among possibilities, relations of the actual to the possible. Since irreducible probability is now a fact of life in physics, and probability is such a modality, there is no escaping this problem. Yet, if we wish to be empiricists, we have nowhere to turn besides thought and language for the locus of possibility. In other words, an empiricist position must entail that the philosophical exploration of modality, even where it occurs in science, is to be part of the theory of meaning.

In sections 1 and 3 I already made clear one important point in the empiricist view of scientific models. They may, without detriment to their function, contain much structure which corresponds to no elements of reality. The part of the model which represents reality includes the representation of actual observable phenomena, and *perhaps* something more, but is explicitly allowed to be only a proper part of the whole model.

This gives us I think the required leeway for a program in the theory of meaning. If the link between language and reality is mediated by models, it may be a very incomplete link – without depriving the language of a complete semantic structure. The idea is that the interpretation of language is not simply an association of a real denotata with grammatical expressions. Instead the interpretation proceeds in two steps. First, certain expressions are assigned values in the family of models and their logical relations derive from relations among these values. Next, reference or denotation is gained indirectly because those model elements *may* correspond to elements of reality. The exploration of modal discourse may then draw largely on structure in the models which outstrips their representation of reality.

A graphic, if somewhat inaccurate way to put this would be: causal and modal discourse describes features of our models, not features of the world. The view of language presented here – that discourse is guided by models or pictures, and that the logic of discourse is constituted by this guidance – I recommend as a general empiricist approach for a theory of meaning without metaphysics.[4]


**Notes**

1. I use the word deliberately: it was a tragedy for philosophers of science to go off on these logico-linguistic tangles, which contributed nothing to the understanding of either science or logic or language. It is still unfortunately necessary to speak polemically about this, because so much philosophy of science is still couched in terminology based on a mistake.
2. The impact of Suppes' innovation is lost if models are defined, as in many standard logic texts, to be partially linguistic entities, each yoked to a particular syntax. Here the models are mathematical structures, called models of a given theory only by virtue of belonging to the class defined to be the models of that theory. See section 7.

3. This answers a question posed in another review, the one by Michael Friedman. Unfortunately Friedman assumed the contrary answer, and built part of his critique on that conjecture.
4. For the background of this paper I wish to refer the reader to pp. 21–30 of the Introduction to F. Suppe (ed.) *The Structure of Scientific Theories* (Urbana: University of Illinois Press, 1974); Ch. 5 'Theories' of R. Giere, *Understanding Scientific Reasoning* (New York: Holt, Rinehart, Winston, 1979); Ch. 3 section 4 of my *An Introduction to the Philosophy of Time and Space* (New York: Random House, 1970); my book *Scientific Image* (Oxford: Oxford University Press, 1980); Patrick Suppes, 'What is a scientific theory?' pp/55–67 in S. Morgenbesser (ed.) *Philosophy of Science Today* (New York: Basic Books, 1967).

## Bibliography

Bell, J.S. (1964), 'On the Einstein Podolsky Rosen Paradox'. *Physics* 1: 195–200.
Bell, J.S. (1966), 'On the Problems of Hidden Variables in Quantum Mechanics'. *Reviews of Modern Physics* 38: 447–452.
Bell, J.S. (1971), 'Introduction to the Hidden Variable Question', in *Foundations of Quantum Mechanics*. Edited by B. d'Espagnat. New York: Academic Press. Pages 171–181.
Beltrametti, E. and van Fraassen B.C. (eds.). (1981), *Current Issues in Quantum Logic*. New York: Plenum.
Cooke, R.M. and Hilgevoord, J. (1980), 'The Algebra of Physical Magnitudes'. *Foundations of Physics* 10: 363–373.
Dalla Chiari, M.L. and Toraldo di Francia G. 'A Logical Analysis of Physical Theories', *Rivista di Nuovo Cimento,* Serie 2, 3 (1973): 1–20.
Dalla Chiara, M.L. 'Formal Analysis of Physical Theories', in G. Toraldo di Francia (ed.) *Problems in the Foundations of Physics.* Amsterdam: North-Holland, 1979).
Davies, E.B. (1970), 'On the Repeated Measurements of Continuous Observables in Quantum Mechanics'. *Journal of Functional Analysis* 6: 318–346.
Davies, E.B. and Lewis J.T. (1970), 'An Operational Approach to Quantum Probability'. *Communications in Mathematical Physics* 17: 239–260.
Edwards, C.M. (1970), 'The Operational Approach to Algebraic Quantum Theory I.' *Communications in Mathematical Physics* 16: 207–230.
Edwards, C.M. (1971), 'Classes of Operations in Quantum Theory'. *Communications in Mathematical Physics* 20: 26–56.
Einstein, A., Podolsky B., and Rosen N. (1935), 'Can Quantum Mechanical Description of Reality be Considered Complete?' *The Physical Review* 47: 777–780.
Foulis, D.J. and Randall C.H. (1974), 'Empirical Logic and Quantum Mechanics'. *Synthese* 29: 81–111.
Glymour, C. (1980), *Theory and Evidence*. Princeton University Press.
Hojman, S., Kuchar K, and Teitelboim C. (1973), 'New Approach to General Relativity'. *Nature of Physical Science* 245: 97–98.
Hojman, S. (1976), 'Geometrodynamics Regained'. *Annals of Physics* 96: 88–135.
Jauch, M.M. (1968), *Foundations of Quantum Mechanics*. New York: Addison-Wesley.
Ludwig, G. (1967), 'Attempt of an Axiomatic Foundation of Quantum Mechanics and More General Theories II'. *Communications in Mathematical Physics.* 4: 331–348.
Marlow, A.R. (1978), 'Quantum Theory and Hilbert Space'. *Journal of Mahematical Physics* 19: 1841–1846.
Marlow, A.R. (1980), 'An Extended Quantum Mechanical Embedding Theorem' in *Quantum Theory and Gravitation*. A. R. Marlow (ed.) New York: Academic Press. pp 71–77.
Millikan, Robert A. (1917), *The Electron*. J. W. M. DuMond. (ed.) Chicago: University of Chicago Press.

Peacocke, C. (1981), 'Not Real But Observable'. Review of van Fraassen (1980). *Times Literary Supplement* 30: 121.

Przelewski, M., *The Logic of Empirical Theories* London: Routledge and Kegan Paul, 1969.

Suppe, F., 'Theories, Their Formulations and the Operational Imperative,' *Synthese* 25 (1972), pp. 129–159.

Suppes, P., 'What is a Scientific Theory?' in *Philosophy of Science Today*. S. Morgenbesser (ed.) New York: Basic Books, pp. 55–67.

Suppes, P. (1974), 'The Structure of Theories and the Analysis of Data'. in *The Structure of Scientific Theories*. F. Suppe (ed.) Urbana: University of Illinois Press. pp. 266–283.

Suppes, P., *Studies in Methodology and Foundations of Science* (Dordrecht: Reidel, 1969).

Van Fraassen, B.C., 'The Charybdis of Realism: Epistemological Implications of Bell's Inequality' *Synthese* 52 (1982), pp. 25–38.

Van Fraassen, B.C., 'A Formal Approach to the Philosophy of Science,' pp. 303–366 in R. Colodny (ed.) *Paradigms and Paradoxes: The Challenge of the Quantum Domain*. Pittsburgh: University of Pittsburgh Press. 1972.

Van Fraassen, B.C. (1980), *The Scientific Image*. Oxford and New York: Oxford University Press.

Van Fraassen, B.C., 'Theory Construction and Experiment: An Empiricist View.' pp. 663–678 in P. Asquith and R. Giere (eds.) *PSA 1980* Vol 2 East Lansing, Mich.: Philosophy of Science Association, 1981.

Weyl, H., 'The Ghost of Modality' in M. Farber (ed.) *Philosophical Essays in Memory of Edmund Husserl*. Cambridge, Mass. 1940.

Wojcicki, R., 'Set Theoretic Representations of Empirical Phenomena', *Journal of Philosophical Logic* 3 (1947), 337–343.

Worrall, J., 'An Unreal Image'. Review of van Fraassen (1980). *British Journal for the Philosophy of Science*. 35 (1984), pp. 65–80.

# The Garden in the machine: Gender relations, the Processes of Science, and Feminist Epistemological Strategies

SANDRA HARDING

*University of Delaware, Newark*

Feminist inquiry in the natural and social sciences has challenged science at three levels.* In the first place, many beliefs claimed to be well-supported by research in biology and the social sciences now appear as androcentric. Thus the processes of inquiry which have supported the androcentric claims no longer appear to be gender-free – and in that sense value-neutral, objective, disinterested, dispassionate, and so forth.[1] In the second place, critics have pointed to constant historical alliances between fledgling sciences and local projects of sexual politics. New sciences have appealed to sexual politics as support for their legitimacy; and men, when threatened by the possibility of shifting social relations between the sexes, have appealed to the new sciences to support the legitimacy of subjugating women. Each has provided moral and political resources for the other.[2] Furthermore, while the processes of inquiry in physics – *the* model of objective inquiry – escape incriminating challenges at the first level, they are not so lucky at the second. The conceptions of nature and inquiry central both to classical and contemporary physics now appear as suspiciously androcentric as do those central to biology and the social sciences.[3] Thus it should not be surprising to find clear signs of androcentrism in heretofore well-supported scientific beliefs. Sexual politics as old as the Garden of Eden appear to have been omnipresent in the purportedly objective 'mechanisms' of scientific inquiry.

My focus here is not these kinds of criticism, but instead feminist justificatory strategies – feminist epistemologies. These justificatory strategies produce a third level of criticism of science, for they challenge the theories of knowledge developed to explain the cognitive legitimacy of science's processes of inquiry. Very different explanations of and prescriptions for the processes of science can be gleaned from these attempts to explain how it is that feminist inquiry can be producing more adequate support for scientific claims than have purportedly value-free processes.

* The issues raised by this paper are discussed more fully in *The Science Question in Feminism* (1986). Research for this essay has been supported by the National Science Foundation, a Mina Shaughnessy Fellowship from the Fund for the Improvement of Post-Secondary Education, a Mellon Fellowship at the Center for Research on Women at Wellesley, and a University of Delaware Faculty Research Grant.

*Nancy J. Nersessian (editor), The Process of Science. ISBN 90-247-3425-8.*

## An epistemological problem for feminism

We can best grasp the depth and extent of the feminist challenge to traditional epistemologies by posing an epistemological problem for feminism, and noticing how feminist inquirers have tried to resolve this problem.

Here is the problem. Feminism is a political movement for social change. Looked at from the perspective of the assumptions of Western science's self-understanding, 'feminist knowledge', 'feminist science', and 'feminist philosophy of science' should be contradictions in terms. Scientific knowledge seeking is supposed to be value-neutral, objective, dispassionate, disinterested, and so forth. It is supposed to be protected from historically specific political interests, goals and desires by the norms of science. These are thought to include commitments to, and well-tested sociological and logical methods for, maintaining rigid separations between the goals of 'special interest' political movements, such as feminism is often perceived to be, and the conduct of scientific research.

And yet many claims made by researchers and theorists in the social sciences and biology, which have clearly been motivated by and produced through guidance by feminist concerns, nevertheless appear more plausible (better supported, more reliable, less false, more likely to be confirmed by evidence, etc.) than the beliefs they would replace. These claims emerging from inquiry which is not only politically motivated, but also guided from beginning to end by political understandings, appear to increase the objectivity of our collective understandings of nature and social life.

Here are five examples of what I take to be the least controversial of the very general conclusions which can be drawn from this research: (1) Women's situation, like men's, is primarily a social matter, not a biological one; (2) Men's understandings are just as gendered as are women's; (3) What the natural and social sciences have found worthy of explanation are primarily aspects of nature and social life which appear problematic mainly from the perspective of men's lives; (4) The design of research and collection and interpretation of evidence in biology and the social sciences frequently is male-biased, thereby hiding from us important aspects of the regularities of nature and social life and their underlying causal tendencies; (5) The more intensely gendered a culture is, the more perverse (distorted, unreliable) will tend to be the dominant (men's) understandings of nature and social life. Each of these very general claims represents a large number of specific claims which have emerged from feminist inspired and guided research in the biological and social sciences, as well as from examinations of the self-understandings of the natural sciences – from the sociology of knowledge.[4] While one need not find any individual one of these claims more plausible than the beliefs it would replace, one must find *some* feminist claim or other more plausible, less false, more likely to be confirmed by evidence, etc. in order to enter the discourse of this essay.

These claims raise fundamental epistemological questions. How can politicized inquiry be increasing the objectivity of inquiry? On what epistemological grounds should these feminist claims be justified? What are the understandings of the processes of science which will account for this apparently bizarre phenomenon of politically motivated and guided research which nevertheless produces more adequate explanations, more objective results of inquiry?

An examination of the reports of feminist research in the natural and social sciences reveals several different epistemological strategies, though two have certainly been more common than the others. It is these two strategies – feminist empiricism and feminist standpoint epistemologies – which I shall discuss, though in conclusion I will draw attention to the 'post modernist' tendencies which should be seen, I suggest, as a caution to the empiricist and standpoint tendencies. Each of these three responses to the epistemological questions has its virtues and its problems. Each also reveals more clearly than have other critiques the problems in the 'mainstream' discourse each draws on. Let us examine these three feminist epistemologies in the order of their increasing threat to the dominant empiricist self-understandings of science.

## Feminist empiricism

The argument here is that it is sexism and androcentrism which are responsible for the false claims made by non-feminist inquirers. Sexism and androcentrism are social biases. They are 'prejudices' based on false beliefs (due to superstition, custom, ignorance, or mis-education) and hostile attitudes. They enter research particularly at the stage of the identification and definition of scientific problems, but also in the design of research and in the collection and interpretation of evidence. They are correctable by stricter adherence to the existing methodological norms of scientific inquiry. Movements for social liberation 'make it possible for people to see the world in an enlarged perspective because they remove the covers and blinders that obscure knowledge and observation'.[5] Thus the women's movement creates the opportunity for such an enlarged perspective. Furthermore, the women's movement creates more female and feminist (male and female) scientists, who are more likely than non-feminist men to notice androcentric biases.

This epistemological strategy is by no means uncontroversial, as one can easily see by looking at the outraged responses to such claims. The ideas that scientific problematics, and the design and interpretation of research have often been male biased, and that women, or politically engaged feminists (male and female), can produce more reliable knowledge claims, are not easily accepted by scientists who have prided themselves on what they take to be rigorous adherence to the norms of inquiry in their particular fields. Nevertheless, it is the least threatening of the feminist epistemological strategies for two reasons. First of all, it appears to leave intact the empiricist understandings of the principles of adequate scientific inquiry which are *de rigeur* for most practicing natural and social scientists. It appears to challenge only the incomplete way empiricism has been practiced, not the norms of empiricism themselves: mainstream science has not rigorously enough adhered to its own norms. Second, one can appeal to historical precedents to increase the plausibility of this kind of claim. After all, wasn't it the bourgeois revolution of the 15th to 17th centuries which made it possible for early modern thinkers to see the world in an enlarged perspective because this great social revolution from feudalism to modernism removed the covers and blinders that obscured medieval knowledge and observation? And wasn't the proletarian revolution of the late 19th century responsible for yet one more leap in the objectivity of knowledge claims as it permitted an understanding of the effects of class relations on social relations and social belief? From this perspective, the contemporary women's revolution, occurring on an international scale, is just the most recent of these modernist revolutions, each of which moves us yet closer to the goals of the creators of modern science and the Western vision science helped to produce.

However, further consideration of feminist empiricism reveals that the feminist component deeply undercuts the assumptions of empiricism in three ways. To borrow a point which has been made about the relationship between Liberal Feminism and its parental Liberal social contract theory, feminist empiricism has a radical future.[6] In the first place, empiricism insists that the social identity of the observer is irrelevant to the 'goodness' of the results of research. It is not supposed to make a difference to the explanatory power, objectivity, etc. of the results of research if the inquirer or his (sic) community of scientists are white or black, Chinese or British, bourgeois or proletarian, male or female in social origin. Scientific method is supposed to be a reliable screen for eliminating the social values of the individual scientist and his historical community of peers from the results of research. But feminist empiricism argues that women (or feminists, male and female) as a group are more likely than men (non-feminists) as a group to produce unbiased, 'objective' results of inquiry. Gender relations are *inside* the processes of science on this account, especially because we can only see one gender directing and practicing science. The feminist component of this strategy undercuts, as usual, the Liberal empiricist assumption that mental states such as beliefs are fundamentally properties of individuals.[7] Empiricism, like the Liberal political philosophy which provides its metaphysical and political foundations, has no conceptual space for recognizing that humans are fundamentally constituted by their position in the relational networks of social life. For empiricism, humans appear as socially isolated individuals who are here and there contingently collected into social 'bundles' we call 'cultures'. Feminist empiricism challenges this metaphysics, epistemology, and politics by asserting that individuals are fundamentally women or men, feminists or (intentional or unintentional) sexists (as well as members of class, racial and cultural groups), and that our 'location' in these various kinds of social relations gives us different potentials for acquiring knowledge. If science were in fact really committed to doing everything possible to maximize the growth of knowledge, feminist empiricism would have it prioritize feminists' (male and female) social experience as the best grounding for the adequacy of scientific claims. Thus a 'feminist physicist' (male or female) is not a contradiction in terms, but the ideal of an objective observer.

Second, a key origin of androcentric bias appears to lie in the selection of problems for inquiry, and in the definition of what is problematic about these phenomena.[8] But empiricism insists that its methodological norms are meant to apply only to the 'context of justification', not to the 'context of discovery' where problematics are identified and defined. We are supposed to consider irrelevant to the process of scientific inquiry whether hypotheses come to one from gazing in a crystal ball, from sun worshipping, from moral or political beliefs, or from the recognition of cognitive problems in the beliefs of one's scientific ancestors and peers. Scientific method will eliminate any 'social biases' as a hypothesis goes through its rigorous tests. Feminist empiricism argues that scientific method is not sufficient to do this; that an androcentric picture of nature and social life emerges from the testing only of hypotheses generated by what men find problematic in the world around them. That is, missing from the set of alternative hypotheses men consider when testing their favored hypotheses are the ones which would most deeply challenge androcentric beliefs, ones which appear plausible only from a feminist understanding of women's distinctive social experience.

Finally, feminist empiricists often in fact should be understood to be arguing that it is precisely following the norms of inquiry as these are formulated in particular scientific disciplines which appears to result in androcentric results of research. The norms themselves have been constructed primarily to answer the kinds of questions men ask about nature and social life, and to prevent scrutiny of the way beliefs which are nearly or completely culture-wide in fact cannot be eliminated from the results of research by the methodological norms of inquiry.[9] A reliable picture of women's worlds and of social relations between the sexes often requires methods of inquiry which are deviant and/or devalued in science. Examples of such 'methods' include, on the one hand, considering the social origins of inquirers and the definitions of problematics to be issues 'inside' the processes of scientific inquiry. On the other hand, they include qualitative methods not appropriate to that purported model of scientific inquiry – physics – and 'deconstructive' methods which require the researcher to critically examine the 'text' of scientific hypotheses, explanations, and practices to discover the social functions and meanings of scientific activity and belief.[10] These deconstructive methods originate in historical inquiry, literary criticism, interpretive traditions in the social sciences, and psychoanalysis. They are methods empiricists are trained to avoid like the plague.

Thus feminist empiricism intensifies recent tendencies in the philosophy and social studies of science to problematize three empiricist assumptions. It questions the assumption that the origin of scientific problems has no serious or long-lasting effect on the results of research. It doubts the potency of empiricist scientific method to maximize objectivity. It challenges the assumed social anonymity of the authors of some of the most widely tested and highly confirmed hypotheses.

### The feminist standpoint

This justificatory approach originates in Hegel's insight into the relationship between the master and the slave, and the development of Hegel's perceptions into the 'proletarian standpoint' by Marx, Engels and Lukacs.[11] The argument here is that human activity, or 'material life', not only structures but sets limits on human understanding. If human activity is structured in fundamentally opposing ways for two different groups, 'one can expect that the vision of each will represent an inversion of the other, and in systems of domination the vision available to the rulers will be both partial and perverse'.[12] Men reserve for themselves the right to perform only certain kinds of human activity, assigning the balance to women and other subjugated groups. What they assign to others they regard as merely 'natural' activity, in contrast to the distinctively cultural activity which they reserve for themselves, though their self-selected activity could not occur unless others were socially assigned to perform the labors they disdain. A feminist standpoint is not an ascribed position, but an achievement which requires at its start that we look at nature and social life from the perspective of that disdained activity – women's social experience – instead of from the partial and perverse perspective available from 'ruling' activities of men.

These theorists have characterized the key features of the incriminating division of human activity by sex/gender in a variety of ways. For Dorothy Smith, it is men's abstract administrative activity vs women's concrete processing of human bodies and of the daily local settings in which all human bodies exist. For Hilary Rose, it is men's exclusively mental and manual labor, vs women's unity of emotional, mental and manual labor – a unity of 'head, hand, and heart'. For Nancy Hartsock, it is men's preoccupation with abstract exchanges, vs. women's 'immersion in the world of use – in concrete, many-qualitied, changing material processes'.[13] The vantage point of women's subsistence activities 'represents an intensification and deepening of the materialist world view and consciousness' available to the male worker in the process of production.[14] For Jane Flax, it is a defensive preoccupation with activity intended to insure rigid separations between self and others, self and nature, the self and it's body vs activity concerned to create reciprocal relations between the self and these various representations of 'the other'. Thus for Flax, feminism represents 'the return of the repressed' infantile social experience to the consciousness of the adult world.[15] These theorizations should probably be understood as complementary rather than as competing representations of the same reality.

Women's social experience provides the possibility of more complete and less perverse human understanding – but only the possibility. Feminism provides the theory and motivation for inquiry, and the direction of political struggle through which increasingly more adequate descriptions and underlying causal tendencies of male domination are revealed. Only through feminist inquiry and struggle can the perspective of women be transformed into a feminist standpoint – a morally and scientifically preferable 'location' from which to observe, explain, and design social life.

The feminist standpoint writings offer a different explanation than did feminist empiricism of the puzzling phenomenon of politically directed inquiry which nevertheless often produces empirically preferable results of research. They identify the 'regularities of social life and their underlying causal tendencies' which account for this better empirical fit. Knowledge is supposed to be based on experience, and the reason the feminist claims can turn out to be scientifically preferable is that they originate in, and are tested against, a less distorting kind of social experience. The feminist standpoint justificatory strategy offers an account alternative to empiricism of the effects of the origins of scientific problems, of the relevance of what science thinks of as 'method' to the adequacy of knowledge claims, and of the identity of the collective authors of scientific beliefs. A complete account of scientific 'method' will begin with principles about which social groups are more likely to select and define the most comprehensive and explanatorily fruitful problematics, principles which recognize the varying merits of collective belief and the origin of such belief in differing social practices.

Like feminist empiricism, the feminist standpoint reveals key problems in its 'parental discourse'. Where marxism insisted sexism was entirely a consequence of class relations, a problem within only the superstructural social institutions and bourgeois ideology, this feminist theoretical tendency sees sexual relations as at least as causal as economic relations in creating forms of social life and belief. Like feminist empiricism, the standpoint approach takes women and men to be fundamentally 'sex classes'. In contrast to marxism, they are not merely or perhaps even primarily members of economic classes, though class, like race and culture, also mediates our opportunities to gain empirically preferable understandings of nature and social life. Depending on whether these standpoint theorists tend toward Socialist Feminism or Radical Feminism, marxist theory is taken to provide only a partial or a fundamentally misguided metaphysics and politics.[16]

The first problem one might raise against this standpoint strategy is a pragmatic one. After all, a justificatory strategy is intended to convince. Those wedded to empiricism, and loath to countenance the possibility that the social identity of the observer can be an important variable in the potential objectivity of the results of research, are likely to find feminist empiricism at least mildly plausible, but the feminist standpoint beyond the pale. One might counter this objection in two ways. One could point out that empiricist practices have been at least moderately successful in shaping inquirers primarily into 'professionals' who are systematically trained to be preoccupied with instrumental rationality and to avoid recognizing the more inclusive components of inquiry processes in science which influence the objectivity of the results of research. However they in fact regard the processes of science, empiricists are trained to justify their activity in mechanical terms. The various 'rational reconstructions' of the growth of science provided by philosophers present science as ideally a mechanism for sorting beliefs. Whether or not individuals are attracted into a scientific career by the vision of knowledge seeking in the service of 'social progress',[17] the socialization procedures in scientific training successfully create inquirers whose understandings of the progressiveness of science are limited to the properties of 'scientific method' itself.[18] Positivism in this original sense of the term – the reduction of the positive effects of science to the properties of scientific method – is alive and well in contemporary scientific self-understandings, and insures that practicing scientists will tend to be identifiable kinds of social persons: ones for whom political activism is perceived as damaging to their objectivity.[19] Furthermore, one might also counter that the novelty of an epistemology provides no guide to its eventual fate. After all, Descartes, Locke, Hume, and Kant were not uncontroversial thinkers, but today their views provide the 'folk wisdom' – certainly not the *critical* theory – of modern, Western cultures. Nevertheless, this pragmatic problem with the standpoint justificatory strategy is a main reason why feminist empiricism has appeared to be a more powerful resource for grounding feminist knowledge claims – in spite of the obvious internal contradictions in that approach indicated earlier.

Additional problems appear from the perspective of the feminist post-modernism to be discussed next. In the first place, can there be 'a' feminist standpoint if women's (feminist) social experience is divided by class, race and culture? The answer to this question appears to be 'yes and no', recapitulating the opposition between Radical and Socialist feminisms. 'Yes', insofar as women have been subjugated, excluded, silenced, and symbolized as 'the other' cross-culturally. Just to be socially created as a women is to be created through at least minimally common material and symbolic social processes in every culture. That common social experience of subjugation, exclusion, silencing, and otherness provides common elements of social experience, even if these elements are only represented in concrete lived experience through differing racial, class and cultural forms. On the other hand, the racial, class and cultural differences between women do matter for a justificatory strategy which insists on the importance of the daily, lived, social experience of women. This problem with the standpoint strategy reflects a similar problem in the Marxist discourse on which it draws. Marx was not much concerned to distinguish between different kinds of proletarian experience. And for other groups subjugated in the 19th century – the victims of imperialism and male dominance – Marx explained why their social experience was not of primary epistemological or revolutionary import-ance. The conditions of women and Africans, like that of serfs and peasants, were a consequence of class subjugation; these groups were to be understood from the perspective of the social experience of the proletariat, not of their own. Late 20th-

Century feminism cannot afford such a cavalier dismissal of the emancipatory potential of difference among the groups subjugated by the hegemony of the white, male, bourgeois, Western world view. A number of contemporary feminist theorists are struggling with how to conceptualize the 'unity across difference' rather than across perceived sameness needed to fine tune the standpoint theory.

A correlative problem appears in the perception of the world assumed by the standpoint approach. As one of the standpoint theorists has herself put the issue:

> perhaps 'reality' can have 'a' structure only from the falsely universalizing perspective of the master. That is, only to the extent that one person or group can dominate the whole, can 'reality' appear to be governed by one set of rules or be constituted by one privileged set of social relations.[20]

Both of these criticisms amount to an accusation that the standpoint approach, like feminist empiricism, is still too firmly lodged within the assumptions of the modern, Western worldview to which scientific rationality is central.

Two comments are in order before we turn to feminist post-modernism. First, both feminist empiricism and the feminist standpoint appear to assert that objectivity never has been and could not be increased by the exclusion or elimination of social values from inquiry – at least not in the cultures in which science has existed and in which it will exist in the forseeable future. Instead, it is commitment to anti-authoritarianism, anti-elitism, anti-domination tendencies which have increased the objectivity of science and will cotinue to do so. The emergence of modern science itself can be accounted for in these terms, as can later leaps in our understandings of nature and social life.[21] Assertions of the objectivity-increasing properties of value-neutrality in fact have been formulated at identifiable moments in history as political strategies to preserve scientific activity from authoritarian tendencies, not from emancipatory ones.[22]

Second, one might be tempted to relativist defenses of feminist claims from either empiricist or standpoint grounds. However, this temptation should be avoided. Feminist inquirers are never saying that sexist and anti-sexist claims are equally plausible – for example, that it's equally plausible to regard women's situation as primarily biological and also as primarily social. It should be noted that relativism appears as an intellectual possibility, and as a 'problem', only for dominating groups at the point where the legitimate hegemony of their views is being challenged. Relativism as an intellectual position or a 'problem' is fundamentally a political issue, not the 'cognitive', or 'logical' issue it is usually conceptualized as in the mainstream philosophy of science and epistemology discourses. As a modern intellectual position, it originated in the belated recognition by ruling groups in modern European cultures that the apparently bizarre beliefs and behaviors of non-Europeans of their own internally subjugated groups had a 'rationality' or 'logic' of their own. At least a minimal commitment to belief in rationality as an inherent human property required the suspicion that the preferred modern, ruling class, Western beliefs and modes of belief sorting might not be the only reasonable ones. Once such minimal skepticism is seriously entertained, of course more terrifying possibilities threaten to rear their ugly heads, and especially in a period of international decline for Euro-American political power. Carried to its most skeptical extremes, such recognitions lead to two pre-dawn nightmares for devotees to empiricist understandings of the processes of science. The first is that the preferred Western beliefs and modes of belief might not even be *as equally empirically supportable* as those of non-Western groups (including women's beliefs in this second category). This first nightmare for Western empiricists then creates the second. The 'best strategies' for achieving empirically reliable belief appear to require fundamental 'real life' commitments to political action on behalf of the emancipation of 'the others' whose voices have been missing from the dominant European discourses of modernity. This possibility undercuts the belief that the progressiveness of science is to be found precisely in the rigid separation of scientific from political processes.

## Feminist post-modernism

Strains of 'post modernism' appear in virtually all of the feminist epistemological writings, cheek by jowl with the empiricist or standpoint tendencies to which post-modernism is opposed in the non-feminist discourses. We have already glimpsed these strains in the criticisms of the standpoint approach which question whether feminism can speak in one Archimedean voice and which are skeptical of the empiricist and standpoint assumptions of a single reality 'out there' for the describing and explaining. This approach is perhaps better thought of as a discourse of justificatory caution rather than as a full fledged attempt to explain or ground feminist knowledge claims. One theorist locates the mainstream origins of post-modernism in such otherwise diverse thinkers as Nietzsche, Derrida, Foucault, Lacan, Rorty, Cavell, Feyerabend, Gadamer, Wittgenstein, and Unger, and in such otherwise disparate discourses as those of semiotics, deconstruction, psychoanalysis, structuralism, archeology/ geneology, and nihilism.[23] These thinkers and movements 'share a profound skepticism regarding universal (or universalizing) claims about the existence, nature and powers of reason, progress, science, language and the 'subject/self'.[24]

Feminist post-modern tendencies create a fruitful strategic dilemma which does not appear in the non-feminist post-modern discourses. On the one hand, can feminists afford to give up the goal of trying to provide 'one, true, feminist story of reality' in the face of the deep alliances between science and the sexist, racist, classist, and cultural domination projects which keep women subjugated? On the other hand, can we afford not to challenge the historically disastrous alliance between knowledge and power?

A hopeful possible opening between the horns of this dilemma can be glimpsed when we note the unusual relationship between the standpoint and post-modernist tendencies within the feminist discourse. The parental post-modern tendency abjures the 'policing of thought' characteristic of mainstream science and epistemology which creates distorted understandings of ourselves and the world around us in service to partisan power politics. Post-modernism and the scientific world view here express the political stance of two different groups within the 'ruling' gender, race, class and culture. It is no accident that for the ruling groups in the West, women's knowledges provide the most intimately known, though deeply repressed, example of 'subjugated knowledges'. In contrast, these tendencies within feminism express internal polarities intended to direct knowledge-seeking by a group not party to the mainstream dispute. Feminist post-modernism, like it's parental discourse, envisions a knowledge-seeking situation where social interests in defensive alliances of knowledge with power no longer exist. The feminist standpoint tendency focusses on how to change the present knowledge-seeking situation so that those interests become anachronistic. Both tendencies appear within the thinking of the very same theorists. Taken together, these tendencies express the perceived importance of pursuing simultaneously the only apparently contradictory goals of creating a feminist story of reality while we struggle toward the dismantling of the gender, race and class dominations which made feminism, a movement for social power, necessary in the first place.

## Conclusion

We appear to have strayed far afield from what is usually meant by 'the processes of science'. But in examining the central epistemological strategies which have been developed to account for the emergence of distinctively feminist scientific research, we are lead to consider the benefits to be gained by regarding all of the social relations of the cultures supporting science and, especially, gender relations – as *inside* the processes of science. We need to eliminate the politics of the garden of Eden which have been constitutive factors in modern science. But in order to do that, we need simultaneously to give up the notion that the processes of modern science can be meaningfully understood in isolation from the processes of daily social relations.[25]

## Notes

1. The literature supporting these claims is immense. For a representative small sampling, see Bleier (1984), Kelly-Gadol (1976), Lowe and Hubbard, eds. (1983), Millman and Kanter, eds. (1975), *Signs* (1975 et seq.), Westkott (1979).

136

2. See, e.g., Bloch and Bloch (1980), Fee (1980), Hall (1973–74), Haraway (1978), Hubbard (1983), Jordanova (1980), Keller (1984), Merchant (1980).
3. See, e.g., Keller (1984), Merchant (1980), Traweek (forthcoming).
4. For samples of this literature, see the citations above; also Harding and Hintikka, eds. (1983).
5. Millman and Kanter (1975), p. vii.
6. Eisenstein (1981).
7. Scheman (1983).
8. See, e.g., Bleier (1984), Keller (1982), Lowe and Hubbard (1983), Millman and Kanter (1975).
9. See sources cited above.
10. For examples of these deconstructive methods in operation, see Bloch and Bloch (1980), Flax (1983), Hubbard (1983), Jordanova (1980), Keller (1983), Keller (1984), Keller and Grontkowski (1983), Merchant (1980).
11. Marx (1964, 1970), Engels (1972), Lukacs (1971). The feminist standpoint epistemologies are developed in Flax (1983), Hartsock (1983a, 1983b), Rose (1983), Smith (1974, 1977, 1979, 1981).
12. Hartsock (1983a), p. 285.
13. Hartsock, (1983a), p. 292.
14. Ibid.
15. Flax (1983), p. 255.
16. For a thorough explication of the differences between Liberal, Marxist, Radical and Socialist feminisms, and criticisms of the first three, see Jaggar (1983).
17. Kuhn (1970).
18. See Chapter 1 of Bernstein (1982) for reports of anxiety by leading social theorists over this problem.
19. Van den Daele (1977), Traweek (forthcoming). This is a generalization with many and notable exceptions to be found in such contemporary groups as Physicians for Social Responsibility, Science for the People, etc.
20. Flax (1984), p. 17.
21. Van den Daele (1978), Zilsel (1942).
22. See Van den Daele (1978).
23. Flax (forthcoming), p. 3.
24. Ibid.
25. This essay was completed in 1984. It has been incorporated, with some changes, into *The Science Question in Feminism* (1986).

## Bibliography

Bernstein, R. (1982), *The Restructuring of Social and Political Theory*, (Philadelphia: University of Pennsylvania Press).
Bleier, R. (1984), *Science and Gender: A Critique of Biology and Its Theories on Women* (New York: Pergamon Press).
Bloch, M. and J. Bloch (1980), 'Women and the Dialectics of Nature in Eighteenth Century French Thought', in C. MacCormack and M. Strathern, eds., *Nature, Culture and Gender* (New York: Cambridge University Press.).
Eisenstein, Z. (1981), *The Radical Future of Liberal Feminism* (New York: Longmans.).
Engels, F. (1972), 'Socialism: Utopian and Scientific', in R. Tucker, ed., *The Marx and Engels Reader* (New York: Norton).
Fee, E. (1980), 'Nineteenth Century Craniology: The Study of the Female Skull', *Bulletin of the History of Medicine* 53.
Flax, J. (1983), 'Political Philosophy and the Patriarchal Unconscious: A Psychoanalytic Perspective on Epistemology and Metaphysics', in Harding and Hintikka (1983).
Flax, J. (1984), 'Gender as a Social Problem: In and For Feminist Theory', paper presented to German Association for American Studies, Berlin.

Hall, D.L. (1973–74), 'Biology, Sex Hormones and Sexism in the 1920's', *The Philosophical Forum* 5.

Haraway, D. (1978), 'Animal Sociology and a Natural Economy of the Body Politic', Parts I and II, *Signs: Journal of Women in Culture and Society* 4: 1.

Harding, S. (1986), *The Science Question in Feminism*. (Ithaca: Cornell University Press).

Harding, S. and M. Hintikka eds. (1983), *Discovering Reality: Feminist Perspectives on Epistemology, Metaphysics, Methodology and Philosophy of Science* (Dordrecht: Reidel Publishing Co.).

Hartsock, N. (1983a), 'The Feminist Standpoint: Developing the Ground for a Specifically Feminist Historical Materialism', in Harding and Hintikka.

Hartsock, N. (1983b), 'Chapter 10', *Money, Sex and Power* (Boston: Northeastern University Press).

Hubbard, R. (1983), 'Have Only Men Evolved?', in Harding and Hintikka.

Jaggar, A. (1983), *Feminist Politics and Human Nature* (Totowa, N. J.: Rowman and Allenheld Publishing Co.).

Jordanova, L.J. (1980), 'Natural Facts: A Historical Perspective on Science and Sexuality', in C. MacCormack and M. Strathern, eds., *Nature, Culture and Gender* (New York: Cambridge University Press).

Keller, E.F. (1982), 'Feminism and Science', *Signs: Journal of Women in Culture and Society: 3.*

Keller, E.F. (1983), 'Gender and Science', in Harding and Hintikka (1983).

Keller, E.F. (1984), *Reflections on Gender and Science* (New Haven: Yale University Press).

Keller, E.F. and C. Grontkowski (1983), 'The Mind's Eye', in Harding and Hintikka (1983).

Kelly-Gadol, J. (1976), 'The Social Relations of the Sexes: Methodological Implications of Women's History', *Signs: Journal of Women in Culture and Society* 1: 4.

Kuhn, T.S. (1970), *The Structure of Scientific Revolutions*, 2nd Edition (Chicago: University of Chicago Press.).

Lowe, M. and R. Hubbard, eds. (1983), *Woman's Nature: Rationalizations of Inequality* (New York: Pergamon Press.).

Lukacs, G. (1971), *History and Class Consciousness* (Cambridge: The MIT Press).

Marx, K. (1964), *Economic and Philosophic Manuscripts of 1844,* Dirk Struik, ed. (New York: International Publishers).

Marx, K. (1970), *The German Ideology,* C. J. Arthur, ed. (New York: International Publishers).

Merchant, C. (1980), *The Death of Nature: Women, Ecology and the Scientific Revolution* (New York: Harper and Row).

Millman, M. and R.M. Kanter, eds. (1975), *Another Voice: Feminist Perspectives on Social Life and Social Science* (New York: Anchor Books).

Rose, H. (1983), 'Hand, Brain and Heart: A Feminist Epistemology for the Natural Sciences', *Signs: Journal of Women in Culture and Society* 9: 1.

Rose, H. and S. Rose, eds. (1979), *Ideology of/in the Natural Sciences (Boston: Schenkman Publishing Co.).*

**Scheman, N. (1983), 'Individualism and the Objects of Psychology', in S. Harding and M. Hintikka (1983).**

*Signs: Journal of Women in Culture and Society* (1975 et seq.), (Chicago, University of Chicago Press).

Smith, D. (1974), 'Women's Perspective as a Radical Critique of Sociology', *Sociological Inquiry* 44.

Smith, D. (1977), 'Some Implications of a Sociology for Women', in *Woman in a Man-Made World,* ed. N. Glazer and H. Waehrer (Chicago: Rand-McNally Publishing Co.).

Smith, D. (1979), 'A Sociology for Women', in *The Prism of Sex: Essays in the Sociology of Knowledge,* ed. J. Sherman and E. T. Beck (Madison: University of Wisconsin Press).

Smith, D. (1981), *The Experienced World as Problematic: A Feminist Method* (Saskatoon: Sorokin Lectures, University of Saskatchewan).

Traweek, S. (forthcoming), *Taking Space and Making Time: The Culture of the Particle Physics Community.*

Van den Daele, W. (1977), 'The Social Construction of Science', *The Social Production of Scientific Knowledge,* E. Mendelsohn, P. Weingart, R. Whitley, eds., (Boston: Reidel Publishing Co.).

Westkott, M. (1979), 'Feminist Criticism of the Social Sciences', *Harvard Educational Review* 49.

Zilsel, E. (1942), 'The Sociological Roots of Science', *American Journal of Sociology* 47.

# The Cognitive Study of Science

RONALD N. GIERE
*Indiana University, Bloomington*

## 1. What might the cognitive study of science be?

The aim of the cognitive study of science is to develop a cognitive *theory* of science. But what is meant by a 'theory' of science? And why do I call it 'cognitive'?[1]

*Why A 'theory' of science?*

'Theory' is a vague and often primarily honorific term that shows up in such diverse linguistic contexts as 'critical theory', 'literary theory', 'the theory of justice', 'number theory', and 'the theory of relativity' or 'evolutionary theory'. When I speak of a 'theory' of science, I mean to imply that the study of science as a cultural enterprise is itself a science. To be sure, it is a *human* science, and that raises questions about the extent to which any human science can be like sciences such as physics or biology. I shall not be diverted by such broad issues here. In general I am optimistic about the similarities, but only time will tell whether this optimism is justified.[2]

*Why 'cognitive'?*

Science is a cognitive activity, which is to say it is concerned with the generation of *knowledge*. Indeed, science is now the major paradigm of a knowledge-seeking enterprise. At the moment, the potentially most powerful resource for studying any cognitive activity is the cluster of disciplines loosely grouped under the heading 'Cognitive Science'. So 'a cognitive theory of science' is intended be a broadly scientific account of science utilizing the resources of the cognitive sciences.[3]

Here one must be careful not to take too narrow a view of what constitutes cognitive science. Some equate it with cognitive psychology. Others would restrict it to the application of computational models of cognition, whether to humans or computers. I would insist on a broad view that includes parts of logic and philosophy, and then runs all the way from cognitive neurobiology, through cognitive psychology and artificial intelligence, to linguistics, and on to cognitive sociology and anthropology. One should not put *a priori* restrictions on what might prove useful in under-

*Nancy J. Nersessian (editor), The Process of Science. ISBN 90-247-3425-8.*
© *1987, Martinus Nijhoff Publishers (Kluwer Academic Publishers), Dordrecht. Printed in the Netherlands.*

standing the phenomenon of modern science.[4]

In advocating a cognitive approach to the study of science as a human activity, I am obviously implying that there are important shortcomings to existing accounts of the nature of science. These existing accounts are to be found primarily in the writings of philosophers and sociologists of science. Of course there is much diversity of opinion within each of these disciplines. Here I will ignore most of this diversity so that the outlines or my own position will be clear.

## 2. Rationality, relativism, and cognition

Philosophical theories of science are generally theories of scientific *rationality*. The scientist of philosophical theory is an ideal type, the ideally rational scientist. The actions of real scientists, when they are considered at all, are measured and evaluated by how well they fulfill the ideal. The *context* of science, whether personal, social, or more broadly cultural, is typically regarded as irrelevant.

Recent sociologists of science, by contrast, emphasize context above all else. Scientific knowledge, it is claimed, is totally relative to context. It is 'contingent'. It is, as one sociologist recently claimed, 'one hundred percent a social construct'.[5]

Of course both philosophers and sociologists have arguments which they use to support and defend their respective views. Here I will only present the views, and explain why I regard them as unsatisfactory. That will provide some background for my subsequent claims that a cognitive approach is more promising.

### *The rationality of science*

This subheading, which repeats the title of a recent survey of the philosophy of science, well expresses the philosophers' concern with *rationality*. The relationship between theory and data, for example, is said to be a 'rational', or even 'logical', relationship. For any given theory, and any given set of data, there is said to be a 'rationally correct' conclusion about the extent to which that data 'rationally supports' the theory. The philosopher's task is to make explicit the principles which scientists intuitively employ in evaluating theories, and to show that these principles are indeed rationally correct.[6]

The major *philosophical* difficulty with this approach has always been demonstrating that a particular principle captures a relationship that is uniquely rational. That is what most of the literature is all about. The inconclusiveness of this literature, after many years of effort, is generally taken to indicate the difficulty of the problem. But it may also be taken as a basis for suspecting that there is something fundamentally mistaken about the whole enterprise.

There is also a severe *empirical* difficulty with the typical philosopher's picture of relations between theory and data. If scientific judgment were indeed guided by intuited principles of rationality, one would expect far more agreement among scientists than in fact exists. Of course, if one's picture of science is gathered mainly from textbooks, then widespread agreement seems to be the order of the day. And indeed it is true that there are large areas of agreement that form the background for current research. But if one looks at any area of *active* research, what one almost invariably finds is individuals and groups with widely divergent, and often passionately defended, opinions. This is the normal state of affairs at the research front.

If there are such things as principles of scientific rationality, then the widespread existence of disagreements in science must be the result of irrational forces or interests. One ends up with a picture of scientific research as a highly irrational activity. But most philosophers begin with the assumption that science is rationality writ large. Clearly something is wrong.

## The social construction of scientific knowledge

Versions of the above title now appear regularly on the covers of books in the sociology, or the sociological history, of science. Science, like the law, is pictured as a thoroughly social construct. Experimental data, on this view, are just one resource among many used in social negotiations over what the content of acceptable theory shall be. So are the traditional scientific virtues like simplicity. In place of the philosopher's principles of rationality one finds only the clash of competing social and professional interests.[7]

The philosopher's charge that any such view leads to *relativism* is welcomed with open arms. Our scientific beliefs about the world are held to be no different in principle than the Azande beliefs about witches. There is said to be no basis other than ethnocentric prejudice for our claims that we are right and the Azande wrong. Indeed, the science of the paranormal could, in different social circumstances, be normal science.[8]

This picture of science at least has the virtue of explaining the almost universal existence of disagreement at the research frontier. Disagreement in science is as natural as disagreement in the halls of Parliament. Indeed, on this view the nature of local disagreements, as well as of background agreement, is fundamentally the same in both science and politics.

The main trouble with this approach to understanding science is that it utterly fails to explain the obvious *success* of science, and particularly the success of science based technology, since the seventeenth century. It makes the success of science and technology similar to the success of liberal democracy. In both cases the success would be explained primarily in terms of changing social relationships.

What such an account fails to explain is how we came to be able to produce insulin in a laboratory or to send instrument packed rockets to photograph Uranus. Such feats require substantial social organization, to be sure, but no amount of purely social organization could have produced such results even as recently a twenty years ago. There must be more going on than changing social relationships.[9]

There is indeed something important missing from the sociological account. That something is not rationality, but causal interaction with the world.

*Cognition and reality*

Again I have taken my heading from a recent book, this time in cognitive psychology. The starting point for cognitive psychology, and the cognitive sciences generally, is that humans have various biologically based cognitive capacities including perception, memory, imagination, and language use. These are employed in everyday interactions with the world. A cognitive theory of science would attempt to explain how scientists use these capacities for interacting with the world as they go about the business of constructing modern science.[10]

Until very recently there has been little study of scientists themselves from a cognitive perspective. So there are as yet no standard cognitive views, for example, of how data and theory are related in science. This is a drawback, but also a blessing. The current shortcomings of both the philosophy and the sociology of science stem from the fact that, in being true to their disciplinary backgrounds, both philosophers and sociologists of science have failed their subject matter – science.

Philosophers since the time of Aristotle have been custodians of the concept of rationality. The possibility that rationality might not be an especially useful concept for understanding modern science is not seriously considered. Similarly, recent sociologists of science have been preoccupied with the idea of building a genuinely social theory of science. Non-social factors are not investigated. Having fewer (or at least less focused) disciplinary commitments, a cognitive theory of science is more free to be true to its subject.

## 3. Representation and judgment

For fifty years philosophers of science have devoted major efforts to elucidating (1) the nature of scientific theories and (2) the criteria for choosing one theory over others. In that time several very different sorts of answers to these questions have been proposed, but the importance of the issues has remained. Recent sociologists of science have concerned themselves less with the nature of theories as such, but have devoted much effort to studying the process by which theories come to be accepted by a majority of a scientific community.

The cognitive sciences do not provide immediate and direct answers to these questions. They do, however, suggest a general orientation and some specific concepts that provide a potentially fruitful background for considering the issues. In the framework of the cognitive sciences, theories would be some sort of *representation,* and the selection of a particular theory as the best available would be a matter of individual *judgment.*

*Representation*

That humans (and animals) have internal *representations* of their environment (as well as of themselves) is probably the central notion in the cognitive sciences. It is the appeal to internal, mental representations, for example, that fundamentally distinguishes cognitive psychology from behaviorism. Depending on the local environment within the cognitive sciences, one finds talk of such things as 'schemata', 'cognitive maps', 'mental models', and 'frames'.[11]

My general view is that scientific theories should be regarded as continuous with the representations studied in the cognitive sciences. There are differences to be sure. Scientific theories are more often described using *written* words or *mathematical* symbols than are the mental models of the lay person. But fundamentally they are the same sort of thing.

Here the only feature of representations I wish to remark is that they are just that – *representations*. To put it baldly, they are like internal maps of the external world. Thus employing the cognitive scientists' framework to think about scientific theories automatically makes one some kind of 'realist'. But it need not commit one to the 'hard' kinds of realism criticized by both philosophers and sociologists.

Schemata are not generally described as being 'true' or 'false', but as 'fitting' the world in limited respects and degrees, and for various purposes. Cognitive scientists are not 'metaphysical realists'. By the same token, there is no suggestion that the world simply presents itself without considerable internal processing or 'interpretation'. The mind (or brain) is not a mirror to nature.[12]

*Judgment*

Philosophers picture scientists as evaluating theories by appeal to some vaguely perceived principles of rational evaluation. Sociologists see a process of social negotiation based on interests and contingent needs. Cognitive scientists are not too different from sociologists. They put more emphasis on particular judgments by *individuals*, and social factors are always filtered through individuals. For both, however, judgment is a simply a natural activity of human agents.

The main difference is that cognitive scientists are willing to investigate the *effectiveness* of an individual's judgmental strategies. They are even inclined themselves to judge some strategies as 'biased'. Sociologists seem unwilling make such judgments. But 'ineffective' or 'biased' is not the same as the philosophers' notion of 'irrational'.[13]

*Categorical versus hypothetical rationality*

When philosophers talk about rationality, they generally mean *categorical* rationality. Thus Aristotle is reputed to have defined humans as rational animals.

Rationality is thus viewed as a *property,* indeed an *essential* property, of being human It unambiguously distinguishes humans from other animals. And it does not come in degrees. An entity either has it or not.

There is another sense of 'rational' which simply means using an effective known means to a desired goal. This is *hypothetical,* or instrumental, rationality, and it does come in degrees. When cognitive scientists investigate peoples' judgmental strategies, they are at most evaluating the instrumental rationality of their subjects. What strategies are subjects using in pursuing their goals? Are they doing as well in achieving their goals as they could in the circumstances? If not, why not? There is no claim that subjects might be 'irrational' in a categorical sense.

## 4. Naturalistic realism

The philosophical fortunes of *scientific realism* have risen and fallen with dramatic rapidity during the past twenty years. In 1965 only a few brave souls defended it. By 1975 it was accepted by a large number, if not a majority, of philosophers of science. By 1985 it was again on the defensive. Yet there has been some progress within the shifting fashions, if only in that the range of alternatives has been extended and clarified.[14]

Here I will simply summarize *scientific realism* as the view that when a scientific theory is accepted, most elements of the theory are taken as representing (in some respects and to some degree) aspects of the world. *Anti-realism* is the view that theories are accepted for their non-representational virtues, such as problem solving effectiveness, or for very limited representational virtues, such as saving the observable phenomena.[15]

Let us call the view that there are rational principles for the evaluation of theories *rationalism.* I realize that this terminology has the misleading consequence that many 'empiricists' turn out to be 'rationalists', but I can think of no better term. *Naturalism,* by contrast, is the view that theories come to be accepted (or not) through natural processes involving both individual judgment and social interaction. No special principles of rational theory choice are involved.

We have, then, a double dichotomy along the dimensions of representation and judgment. The result is four possible combinations of views: Rational realism, rational anti-realism, natural realism, and natural anti-realism.

Philosophers of science split over realism about theories, but tend strongly to be rationalists about theory choice. Recent sociologists of science are unanimously both anti-realists about theories and naturalists about theory choice. Partly this is because sociologists tend not to distinguish the two issues of representation and judgment. In much of the sociological literature, 'rationalism' and 'realism' are synonymous, making 'naturalistic realism' a contradiction in terms.[16]

The fourth possibility, *naturalistic realism,* is the obvious place to house a cognitive theory of science. Not that there are not some occupants already there, including some important philosophers of science. But the location has yet to achieve the prominence it deserves. Locating the cognitive study of science in this space will hopefully help raise the respectability of the neighborhood.[17]

## 5. Can the philosophy of science be naturalized?

Naturalistic realism is part of a more general *naturalism* – the view that all human activities are to be understood as entirely natural phenomena like the activities of chemicals or animals. It is also part of a long tradition of using science in the attempt to understand science itself. Like any such general viewpoints, naturalism and the scientific study of science have been 'refuted' by philosophers dozens of times. What makes naturalism attractive now, however, is not that such philosophical arguments have been (or even can be) refuted, but the growing *empirical success* of the cognitive sciences. This is the historical pattern. Proponents of the new physics of the seventeenth century won out not because they explicitly refuted the arguments of the scholastics, but because the empirical success of their science rendered such arguments irrelevant.

Nevertheless, it may help eliminate some confusions and misunderstandings if we consider a few of the contemporary philosophical arguments *against* naturalism. These have mainly been provoked by W.V.O. Quine's spirited advocacy of 'naturalized epistemology' nearly two decades ago. Many philosophers regard the philosophy of science as a special case of epistemology. To naturalize the philosophy of science is thus to naturalize a part of epistemology as well.[18]

*Must a naturalistic philosophy of science define rationality?*

Among the most ardent opponents of naturalized epistemology has been Quine's Harvard colleague, Hilary Putnam. One of Putnam's arguments, applied to a cognitive theory of science, is this. A cognitive theory of science would be committed to a *definition* of rationality of the form: A belief is rational if and only if it is acquired by employing some specified cognitive capacities. But any such formula is either obviously mistaken or vacuous, depending on how we restrict the range of beliefs to which the definition applies. If it is meant to cover *all* beliefs, then it is obviously mistaken because some people clearly do acquire some irrational beliefs using the same cognitive capacities as everyone else. However, restricting the definition to *rational* beliefs renders the definition vacuous. So the program of constructing a naturalistic theory of science along cognitive lines is a non-starter.[19]

The reply is that a naturalistic theory of science need not be committed to any such definition. It is clear that Putman is assuming a categorical conception of rationality. The naturalist in epistemology denies that such a conception can be given any coherent content. For a naturalist there is only hypothetical rationality. Indeed, many naturalists, myself included, would prefer to speak simply about effective goal directed action and drop the word 'rationality' altogether. So Putnam's insistence that the naturalist in epistemology must provide a definition of categorical rationality in naturalistic terms is just a disguised way of begging the question – of insisting that there must be a coherent concept of categorical rationality.

*Can instrumental rationality be enough?*

One way to challenge a naturalist approach is to argue that a rationality of *means* is not enough. There must be a rationality of *goals* as well. In support of this view it has been argued there is no such thing as rational action in pursuit of an irrational goal.[20]

This sort of argument gains its plausibility mainly from the way philosophers use the vocabulary of 'rationality'. If one simply drops this vocabulary, the point vanishes. Clearly there can be effective activity in pursuit of any goal whatsoever – as exhibited by the case of the proverbial efficient Nazi. The claimed connection between instrumental and categorical rationality simply does not exist.

Moreover, investigating the actual goals of any group of scientists is an empirical matter, as is the investigation of the effectiveness of their means. That there is any way of evaluating the 'rationality' of these goals, apart from considering other goals to which scientific activities might be an efficient means, is problematic. That such evaluation *must* be possible is an unproven philosophical article of faith.

*Rationality and the ethics of belief*

At one point Putnam declares that any naturalistic epistemology would eliminate *normative* reason, which is both 'immanent' and 'transcendent'. But to do so, he claims, would be to commit 'mental suicide'. He goes on to claim that reason requires language, which involves reference, which involves *values*. And values cannot be eliminated.[21]

I would not pretend to be able to unravel the threads of these arguments. But I do think one can see in a general way what is at issue. Putnam believes that there is a categorical difference between what scientists in fact believe and what they *rationally ought* to believe, just as there is often claimed to be a categorical difference between what people do and what they *morally ought* to do. 'Ought' cannot be eliminated or otherwise reduced to 'is'.

Now there is one thing right about this. Epistemological naturalism and ethical naturalism go together. One can accept both or reject both, but one cannot have one without the other. For what seem to me mainly historical reasons, philosophers tend to think it is easier to accept epistemological naturalism than its ethical sibling. Thus the rejection of ethical naturalism is used to support the rejection of epistemological naturalism. For me the relationship has always seemed to go the other way. The obviousness of ethical naturalism supports epistemological naturalism.

I agree, however, that Putnam is following a long philosophical tradition, one which can not easily be refuted by argument alone. And I can claim no special expertise in the study of ethics. There is just one further general line of objection to naturalistic philosophies of science I wish to consider.

## 6. Must a naturalistic study of science be viciously circular?

There is another long standing objection to eliminating traditional epistemological questions in favor of questions about effective means to desired goals. To show that some methods are effective one must be able to show that they can reach the goal. And this requires being able to say what it is like to reach the goal. But the goal in science is usually taken to be something like 'true' or 'correct' theories. And the traditional epistemological problem has always been to *justify* the claim that one has in fact found a correct theory. Any naturalistic theory of science which appeals only to effective means to the goal of discovering correct theories must beg this question. Thus a naturalistic philosophy of science can be supported only by a circular argument that assumes some means to the goal are in fact effective.

Before proceeding any further it is worth noting that, apart from some philosophers and sociologists of science, there is practically no knowledgeable group in the whole of western (if not world) culture that seriously questions the obvious successes of science. Who would seriously question that Watson and Crick discovered the basic structure of DNA in the early 1950s? Or that geologists in the 1960s discovered that the crust of the earth consists of shifting plates? This is not to say that what most educated people believe is therefore true. But the disparity should give one pause. Why is there so great a gulf between professional students of the nature of science and the educated public?

In the case of philosophers of science the answer must be sought in the history of what is still called 'modern' philosophy – Descartes' seventeenth century program of universal doubt.

### The Cartesian circle

Descartes set out to provide a firm foundation for the developing science of his time by first doubting everything he could. In the course of rescuing himself from this self-created predicament, he argued for the existence of God on the basis of having a clear and distinct idea of a perfect being which necessarily exhibits the perfection of existence. Later he attempted to justify his appeal to clear and distinct ideas by arguing that a perfect being would not allow one to be deceived by a clear and distinct idea. But this, his critics objected, is 'circular'. The objective is to give a 'straight' argument from some indubitable premises to the desired conclusion.

Before the scientific revolution it was among the main goals of philosophy to justify theology. Many later philosophers concluded that science is better than theology and took it upon themselves to justify science. The object to be legitimated changed, but the goal and the methods remained the same: that of providing a non-circular argument from secure premises. What is surprising is that so many people who would reject the task of providing this sort of justification for theology nevertheless thought that the same goal might be achieved for science.

The enterprise of trying to justify science without appeal to any even minimally scientific premises has been going on without conspicuous success for three hundred years. One begins to suspect the lack of success is due to the impossibility of the task. Perhaps there just is no place totally outside science from which to justify science. At the very least one might conclude that the task is not going to be accomplished any time soon by ordinary mortals. I am willing to take this as sufficient grounds for an ordinary mortal to try something else.

*Is a naturalized philosophy of science possible?*

It begins to look as if the answer to this question depends on how one *defines* the philosophy of science. If one adopts the standpoint of traditional epistemology, then clearly a naturalized philosophy of science is impossible. Traditional epistemology insists on solving the Cartesian riddle – refuting universal skepticism – by establishing a categorical conception of rationality. Nothing less will do. To pursue a naturalized philosophy of science is to abandon this objective. If, on the other hand, one takes a broader view of the philosophy of science as part of the enterprise of developing some theoretical understanding of the nature of science, then a naturalized philosophy of science is clearly possible.[22]

But one should not be deceived into thinking that the issue is merely a matter of definition. At stake is the *autonomy* of the philosophy of science and its assumed status as custodian and arbiter of scientific rationality. If the philosophy of science is naturalized, philosophers of science will find themselves on equal footing with psychologists, sociologists, and others for whom the study of science is itself a scientific enterprise. The most status philosophers of science could claim is that of the *theoreticians* of a developing cognitive science of science, on the model of theoretical, as opposed to experimental, physics. I am inclined to think that would be status enough.[23]

## 7. Evolutionary naturalism

If one seeks a 'deeper' foundation for the cognitive study of science, that can be provided. It lies not in epistemology, nor in the philosophy of language, but in evolutionary theory.

Human perceptual and other cognitive capacities have evolved along with human bodies. We share many of these capacities with other primates and even lower mammals. Indeed, those parts of our brains responsible for our more advanced linguistic abilities are built upon and linked to those parts which we share with other mammals. There can be no denying that these capacities are fairly well-adapted to the environment in which they evolved. Without considerable adaptation, we would not be here. Nor are these capacities trivial. The perceptual and neural processing required just for a human to walk without falling or bumping into things is fantastically complex.

Empiricist philosophers emphasized the role of immediate perceptual experience in their analyses of knowledge because of the high degree of subjective certainty attached to such experience. Their problem was then to get beyond the immediate perceptions. From an evolutionary perspective, the subjective certainty of perception is indeed causally connected with the source of the reliability of such judgments, which lies in our evolved capacities for interacting with our world. But the operation of these capacities is largely unrecorded in our conscious experience. Nor is there any obvious subjective trace of the evolutionary process that produced these capacities. Indeed, the biological nature and evolutionary origin of the connection between our subjective experience and our cognitive capacities is an interesting, and largely unsolved, scientific problem. So is the mapping of particular perceptual and cognitive capacities with particular types of conscious experience. The general *reliability* of the mechanisms that generate many of our perceptual and cognitive judgments, however, is, from an evolutionary perspective, not open to serious question.

Rationalist philosophers, on the other hand, focused on our more general subjective intuitions, such as that space has three dimensions and that time exhibits a linear structure. These judgments seem to be built into the way we think. And indeed they are, for the aspects of the world relevant to our biological fitness have roughly that structure. But rationalists, like empiricists, could not see how to get beyond their subjective intuitions. They ended up in the ironic position of denying that the world (in which our capacities in fact evolved) is knowable by humans (that is, organisms possessing those evolved capacities). In proclaiming the subjective structure of our experience a necessary condition for all knowledge they ruled out the possibility that we should ever be able to discover that the structure of the world as a whole in fact differs from that revealed by our intuitions – that, for example, space and time are really a four-dimensional space-time.

Thinkers struggling to understand the nature of their own knowledge in the seventeenth and eighteenth centuries may be forgiven for not appreciating the evoluationary point of view. A century after Darwin, a similar lack of appreciation is less forgivable.

At this point I am tempted to wax Wittgensteinian and suggest that traditional epistemology has been in the grip of a powerful picture, that of the 'straight line' justification. It is this picture that makes the standard charge of 'circularity' so powerful. Evolutionary theory provides an alternative picture. By looking back at evolutionary history through the lens of evolutionary theory, scientists can better understand their own cognitive situation and investigate their own cognitive capacities. What seem to the traditional epistemologist like vicious circles are, in this alternative picture, 'positive feedback loops'. The existence of such loops is not a defect, but one of the things that makes modern science so powerful.

## Can sociobiology help?

The capacities evolution favors, of course, are just those that confer biological fitness, that is, the ability to survive and leave offspring. The ability to do modern science, one might conclude, had nothing to do with the evolution of our perceptual and cognitive capacities. The general problem faced by a naturalistic philosophy of science, then, is to explain how creatures with our natural endowments manage to learn so much about the detailed structure of the world – 'of atoms, stars, and nebulae, of entropy and genes.'[24]

Some sociobiologists have recently suggested that this task may be easier that one might at first think. They suggest that humans evolved capacities, described by 'epigenetic rules', for making typically scientific inferences. Moving from the observation of repeated 'coincidences' to the supposition that there is a 'common cause', is one such suggested rule.[25]

The existence of such evolved capacities is clearly an empirical hypothesis, though one for which evidence either for or against is difficult to acquire. Such capacities, however, seem not to be required by a cognitive theory of science. Accordingly I shall make no appeals in this direction.

## Evolutionary epistemology

It should be noted that my own appeal to evolutionary theory is far more modest than that of numerous advocates of 'evolutionary epistemology'. It is limited to providing an alternative 'foundation' for the study of science. It explains why the traditional projects of epistemology, whether in their Cartesian, Humean, or Kantian form, were misguided. And it shows why we should not fear the charge of circularity.

Evolutionary epistemologists typically advance the much more ambitious thesis that evolutionary theory provides a good model for the overall development of scientific knowledge. A strictly Darwinian account, for example, would try to show how scientific knowledge evolves through some mechanism of random variation and selective retention.[26]

Whether evolutionary models fit the development of science itself in interesting ways must be an empirical matter. I would not at this point take a strong stand one way or the other. But I myself do not approach the study of science with evolutionary models in hand, hoping to make them fit somehow or other. Nevertheless, I do appreciate the power of evolutionary models and am quite willing to be guided by evolutionary concepts where they seem fruitful.

*Does evolutionary naturalism imply a social theory of science?*

I have claimed that evolutionary naturalism supplies an empirical foundation for the cognitive study of science, which in turn is incompatible with a thoroughgoing social theory of science. But some social theorists would argue the reverse. Evolutionary naturalism, they claim, requires a social theory of science. The argument can be given either a historical or contemporary form.[27]

The historical form is this. The evolved cognitive capacities of humans were in place long before the scientific revolution, and did not change significantly between, say, 1400 and 1700. It therefore cannot be these capacities that made the difference. It must have been 'higher level' social factors.

The contemporary version would be this. Imagine two research groups with strongly opposing theories. Surely the range of cognitive abilities in the two groups does not differ significantly. And they both have access to the same evidence. So what explains the dispute must be social in nature.

The general form of reply is this. The mere *existence* of evolved cognitive capacities does not determine the *uses* to which they are put. That depends on many things, including, of course, the social context. So while one may need to invoke social factors to explain why scientists use their cognitive capacities in some ways rather than others, one must still appeal to their cognitive abilities to understand what they are doing when they employ these abilities in the specified ways.

There are surely crucial social factors in the story of how Galileo acquired his mathematical knowledge, his skill at building telescopes, and his motivation to use both, when he discovered the moons of Jupiter. But the way he used these abilities is essential to any explanation of how he was able to make this discovery.

Similarly, the two competing research groups may not differ in basic cognitive abilities, but they probably do in the kinds of models with which they are comfortable. They may also differ in their experimental skills with different sorts of equipment. Such differing uses of their cognitive abilities would surely be an important part of any explanation of why they disagree which model is best.

In sum, a cognitive approach to the study of science does not reject social factors as being important for understanding science. It only denies that they are everything.

## 8. What might a cognitive theory of science be like?

Developing a scientific theory of science is a *reflexive* enterprise in the sense that one is practicing a form of the very kind of activity under study. Thus one necessarily begins with some commitments about one's subject embedded in one's own practice. This need not lead to paradox or irredeemable bias so long as one is able to change one's practice in the light of one's own findings. But because the mechanisms of self-correction are sure to be fairly inefficient, it is important that one begin with a good first approximation.

One common conception of science is as the search for 'laws' in the form of generalizations. My view is that such a conception is so far from correct that it practically dooms the enterprise of constructing a theory of science to triviality. Two examples, one each from sociology and philosophy, will help make this point.

At the end of a classic sociological study of the development of radio astronomy in Britain, the authors set out to compare their study with five other studies. They proceed to list fifteen 'factors in scientific innovation and specialty development' which are exhibited in at least one of the resulting half dozen studies. Several of these factors are exhibited in all six of the studies surveyed. The factors cited include such things as 'marginal innovation' (introduction of novel ideas or techniques by people at the margin of the profession) 'mobility' (into or out of related specialties), and 'creation of new journal'. Although they do not describe their remarks this way, these sociologists are clearly looking for some 'laws' of innovation and specialty development such as: 'The formation of a new specialty tends to be accompanied by the creation of a new journal'.[28]

In a similar vein, a group of historically oriented philosophers of science has recently embarked on a project of 'testing methodologies against history' – by which they mean comparing various recent claims about the development of science with historical cases. An example of the roughly 250 claims to be tested is: 'A set of guiding assumptions is never rejected unless an alternative set is available'.[29]

This is one place where evolutionary theory provides a useful model. In the nineteenth century, some biologists looked for 'laws of evolution' among existing species and in the fossil record. Bergmann's Law, for example, states that for species of warm blooded vertebrates, races living in cooler climates are larger than races living in warmer climates. The trouble with such laws is that they hover between falsity and vacuity. If stated with sufficient precision to be informative, there are always some exceptions, e.g., burrowing mammals. If stated with sufficient generality or qualification so as to avoid exceptions, they cease to be usefully informative.[30]

The reason such 'laws' have so dubious a status is that they are attempts to generalize over what are partly *initial conditions* (including the environment) of evolving systems of organisms. At best such statements express rough statistical generalizations reflecting the average selective forces of different environments on similar populations of organisms. Even so, they provide no insight whatever into the *mechanisms* of evolutionary development.

The 'laws' of scientific development noted above are like the nineteenth century laws of evolutionary development. They provide at most some rough generalizations covering a very restricted class of cases. The problem is that the phenomenon of science is too variable at the level of the types of factors surveyed. Just as nature may solve an evolutionary problem in many different ways, so scientific development may occur in many different ways. A theory of science, like the modern theory of evolution, must be based on models which capture the 'deeper structure' of the subject matter.

Standard evolutionary models have three components: (i) random variation of traits among members of a population, (ii) differential fitness, relative to the environment, of organisms with different traits, and (iii) inheritance of traits. Any population of organisms satisfying these conditions will evolve, provided only that the environment is not so hostile as to preclude survival. On this characterization, evolutionary theory does not tell us much about how specific populations will evolve in particular environments. But it tells us much about the process of evolution.

Using the notions of representation and judgment, one could begin to develop an analogous model for science. Variations in representations exist within any population of scientists. Scientific judgment acts differentially on different representations. Representations are passed on to later generations through teaching and apprenticeship.

As indicated earlier, it is an open question how fruitful it might be to develop this analogy in greater detail. At the moment, the more promising course for the cognitive study of science would seem to be investigating the roles played by the specific cognitive mechanisms of representation and judgment in scientific research.

## 9. A role for history?

Up to now I have not mentioned the history of science. This may seem surprising because of all the disciplines comprising Science and Technology Studies, the history of science is currently the most active and in many ways the most exciting. But historians offer us no explicit *theory* of science. To be sure, historians *have* theories of science, often borrowed from elsewhere, which are implicit in their practice. But these theories are not the focus of their inquiry.

Within the history of science during the last generation there has been a steady shift away from viewing the history of science as a branch of intellectual history toward seeing it more as part of social history. In this, historians of science are partly just following a trend within the broader discipline of history itself. In some cases, historians have been explicitly inspired by suggestions from the sociology of science. But generally it is the sociologists who provide the theoretical formulations of what a social theory of science should be like.

What, then, is the role of history in the enterprise of constructing a theory of science? For philosophers of science concerned to defend a categorical theory of scientific rationality, the role has always been problematical. History provides us only *descriptions* of how science has been pursued. From descriptions one cannot derive substantive *norms* governing how science should be pursued. History, in this framework, can at best provide 'illustrations' of science having been pursued as some proposed norms would prescribe.

Any naturalistic theory of science opens the door to the possibility suggested by Kuhn in the very first chapter of *The Structure of Scientific Revolutions*. The history of science, he said can provide empirical evidence against which naturalistic theories of science, like any scientific theories, can be judged. But there are pitfalls as well as promises in this suggestion. In particular, much depends on what kind of a theory of science one seeks.

*Are there laws of scientific development?*

The usual interpretation has been that history provides evidence relevant to theories of scientific *development*. This interpretation has been reinforced by the fact that Kuhn himself proposed an account of scientific development: preparadigm science, normal science, crisis, revolution, new normal science, and so on. But can any so general a theory of scientific development by very enlightening? Is it worth looking for 'laws' of scientific development?

The example of economics is sobering. The most successful theories in economics, whether micro or macro, are *equilibrium* theories. They provide no dynamics for development. Grand developmental theories, like those of Marx, are in disrepute – at least in the West, and increasingly in the East as well. But even lesser, capitalist, 'stage' theories of economic development are not well regarded. Why should a general theory of scientific development fare any better?

In biology, research on the development of *individuals* has just recently come back into vogue. Not, however, in the form of laws of development, but in the form of genetic and molecular mechanisms which produce development. This follows the pattern in evolutionary theory as well. Molecular genetics provides the basic mechanisms which produce evolution in populations. But here again there are no resulting 'laws' describing developmental stages. What the genetic mechanisms actually produce depends on environmental conditions which evolutionary theory itself does not describe.

*The mechanisms of scientific change*

If we take the cognitive mechanisms of representation and judgment as the rough analogues of genetic mechanisms, then we get a very different picture of relations between theories of science and the history of science. They are something like that between the genetical theory of evolution and the fossil record. The fossil record is nicely interpretable in terms of evolutionarry models, and the fact that it is so interpretable provides some basis for taking these models to be correct. But evolutionary processes are even better studied in the laboratory using contemporary organisms and the modern techniques of molecular biology.

It is natural that we humans should be particularly interested in that part of the fossil record relevant to the early development of our own species. And, indeed, many characteristics of contemporary humans have no other explanation than that they developed in the particular circumstances confronting our early ancestors. But to understand the life of contemporary humans it is necessary to study contemporary humans.

In like manner, the history of science naturally holds a particular fascination for those interested in contemporary science. And, indeed, some characteristics of current science no doubt have no other explanation than than they developed in the particular historical circumstances of earlier science. But to understand the role of science in contemporary culture, there is no substitute for studying the workings of contemporary science.

Another strong trend among contemporary historians of science is increasing attention to twentieth century science. At some point, of course, the history of science blends together with the study of contemporary science. At this point the relationship between historians and theorists of science becomes analogous to that between experimentalists and theorists in other areas of science. Or, because the study of science is more like sociology or anthropology than like physics or biology, the relationship becomes similar to that between those who gather data in the field and those who use that data to develop and test their theories.

## Conclusion

A cognitive approach to the study of science has the potential for resolving differences between philosophers and sociologists of science, as well as differences within each of these fields. And this potential will increase with further developments in the cognitive sciences. The ultimate test of a cognitive approach, however, is how useful it is in developing a detailed understanding of the phenomenon of modern science.

## Notes

1. This paper is adapted from the opening chapter of *Explaining Science: A Cognitive Approach* to be published by the University of Chicago Press. The support of the National Science Foundation through its program of Scholars Awards in History and Philosophy of Science is hereby gratefully acknowledged.

2. It will already be evident that I shall not be concerned with what in Continental circles is called the 'critique' of science. In my admittedly limited experience, such critique too often takes place in glaring ignorance of the nature of scientific activity. I am thinking in particular of Habermas' views, as expounded, for example, in *Knowledge and Human Interests* (1971), and discussed in Thomas McCarthy's *The Critical Theory of Jurgen Habermas* (1978).

3. The idea of a 'cognitive' theory of science is to be found in a number of recent works including *Scientific Thinking* (1981), edited by Ryan Tweney, Michael Doherty, and Clifford Mynatt; *The Cognitive Paradigm* (1982), by Marc DeMey; and *Science as Cognitive Process* (1984), by Robert Rubinstein, Charles Laughlin, and John McManus; It is clearly foreshadowed in earlier works by Donald Campbell (1959, 1974), Herbert Simon (1957), Howard Gruber (1974) and, of course, Jean Piaget (Gruber and Voneche, 1977). It can also be found implicitly in Kuhn (1962), and especially in his 'Second Thoughts on Paradigms' (1974).

   Many philosophers and philosophers of science have recently turned their attention toward the cognitive sciences with the intent of applying philosophical concepts and methods to investigate these sciences. It is essential to realize that the intent of a cognitive theory of science is just the *reverse:* to use the concepts and methods of the cognitive sciences to study science itself.

4. A broad view of cognitive science is promoted, for example, by Howard Gardner in his popular, but very informative, *The Mind's New Science* (1985). This is the best overall introduction to the field currently available. For a somewhat narrower view, more centered on artificial intelligence, see the introduction and essays in *Mind Design* (1981), edited by John Haugeland. For a still narrower view see Zenon Pylyshyn's *Computation and Cognition* (1984).

5. The sociologist in question is Andrew Pickering, and the quotation is not from his very substantial book (1984), but from a recent oral presentation.

6. The subtitle comes from William Newton-Smith's (1981) recent book. This characterization of relations between data and theory better fits logical empiricist or later 'analytical' studies than it does more 'historical' studies in the philosophy of science. But those historically minded philosophers of science who speak rather of 'research programmes' (Lakatos, 1970) or 'research traditions' (Laudan, 1977) are equally concerned with rationality. For present purposes, then, the otherwise great differences between these two schools within the philosophy of science can be ignored.

7. Here I am thinking both of the 'Edinburgh School', as represented by Barnes (1974), Bloor (1976), Barnes and Bloor (1982), Barnes and Edge (1982), and Shapin (1982), and of more ethnographic studies like those of Latour and Woolgar (1979), Knorr-Cetina (1981), and Knorr-Cetina and Mulkay (1983).

8. The explicit target of these remarks is Collins and Pinch's *Frames of Meaning* (1982).

9. The claim that constructivist sociological accounts fail to explain the success of science has been developed by Laudan (1984).

10. The subtitle follows Neisser (1976).

11. The notion of 'schemata' goes back almost a half century to the work of Piaget (Gruber and Voneche, 1977) and Bartlett (1932). E. C. Tolman (1948) talked about 'cognitive maps' in both animals and humans. For a more recent discussion, see Chapter 6 of Neisser (1976). Johnson-Laird (1983) has popularized the term 'mental model' within cognitive psychology while Minsky (1975) introduced the term 'frame' into the computer science literature.

12. The references here are to Putnam (1982) and Rorty (1979).

13. See Nisbett and Ross (1980); Kahneman, Slovic, and Tversky (1982); and Faust (1984).

14. For the most recent presentations of views on realism see the essays in Leplin (1984), and Churchland and Hooker (1985).

15. The references here are to Laudan (1977) and van Fraassen (1980) respectively.

16. The combination of anti-realism regarding theories and naturalism regarding scientific judgment matches pretty well with what most people call 'relativism'. In the literature, the contrasts tend to be dichotomous, as in Hollis and Lukes' *Rationality and Relativism* (1982) and Laudan's (1984) discussion of realism and relativism. Failure to sort out the two dimensions, representation and judgment, leads to confusion, as when Laudan suggests that realism and relativism may be 'orthogonal', but then pronounces them genuine 'contraries'.

17. Campbell (1966) has been a naturalistic realist for years. Among others who seem clearly to be in this neighborhood are Boyd (1981), Churchland (1979), and Hooker (1978).

18. For an up to date introduction to the literature on naturalistic epistemology see Kornblith (1985).
19. Putnam (1982, p. 5).
20. Siegel (1985).
21. Putnam (1982, p. 22).
22. The case for a narrower, non-naturalistic, view of epistemology is well argued by Stroud (1981), reprinted in Kornblith (1985).
23. The traditional philosopher's attitude toward the actual practice of science is illustrated by the following remark of Nelson Goodman's (1972, p. 168), quoted approvingly by Siegel (1980).

    The scientist may use platonistic class constructions, complex numbers, divination by inspection of entrails, or any claptrappery that he thinks may help him get the results he wants. But what he produces then becomes raw material for the philosopher, whose task is to make sense of all this: to clarify, simplify, explain, interpret in understandable terms. The practical scientist does the business but the philosopher keeps the books.

    Goodman is extreme only in his arrogance.
24. The phrase is borrowed from George Gamow (1947) who adapted it from a famous line in Lewis Carroll's *Through the Looking Glass*.
25. See Ruse (1986). The notion of an epigenetic rule has been developed by Lumsden and Wilson (1981).
26. See Campbell (1960, 1974), Toulmin (1972), and the essays in Plotkin (1982). For an overview of evolutionary epistemologies, see Bradie (1986).
27. I learned this argument from Bloor in private conversation.
28. Edge and Mulkay (1976, p. 382).
29. I refer to the program of Laudan and his colleagues (1986).
30. One can imagine a hypothetical Darwin setting out to create a theory of evolution by gathering together all such claims in the known literature and then attempting to test them against the empirical record. Such a Darwin would be as unknown today as is Bergmann. One hundred such 'rules' are to be found in Berhnard Rensch's *Biophilosophy* (1971, pp. 131–139). This reference I owe to David Hull who generously loaned me his copy.

## Bibliography

Barnes, Barry (1974), *Scientific Knowledge and Sociological Theory*. (London: Routledge & Kegan Paul).

Barnes, Barry and David Edge, eds. (1982), *Science in Context*. (Cambridge, MA: The MIT Press).

Barnes, Barry and David Bloor. (1982), 'Relativitism, Rationalism and the Sociology of Knowledge'. In M. Hollis and S. Lukes, eds., *Rationality and Relativism*. (Cambridge, MA: The MIT Press, 21–47).

Bartlett, F.C. (1932), *Remembering*. (Cambridge, England: Cambridge University Press).

Bloor, David (1976), *Knowledge and Social Imagery*. (London: Routledge & Kegan Paul).

Boyd, R. (1981), 'Scientific Realism and Naturalistic Epistemology'. In P. D. Asquith and R. N. Giere, eds. *PSA 1980,* Vol. 2. (East Lansing, MI: The Philosophy of Science Association).

Bradie, M. (1986), 'Assessing Evolutionary Epistemologies'. *Biology and Philosophy,* 1.

Campbell, D.T. (1959), 'Methodological Suggestions from a Comparative Psychology of Knowledge Processes'. *Inquiry,* 2, 152–82.

Campbell, D.T. (1960), 'Blind Variation and Selective Retention in Creative Thought as in Other Knowledge Processes'. *Psychological Review,* 67, 380–400.

Campbell, D.T. (1966), 'Pattern Matching as Essential in Distal Knowing'. In *The Psychology of Egon Brunswick,* K.R. Hammond, ed. (New York: Holt, Rinehart, and Winston) 81–106.

Campbell, D.T. (1974), 'Evolutionary Epistemology'. In P. A. Schilpp, ed., *The Philosophy of Karl Popper*. (La Salle: Open Court) 413–463.

158

Churchland, P.M. (1979), *Scientific Realism and the Plasticity of Mind*. (Cambridge: Cambridge University Press).

Churchland, P.M. and C.A. Hooker. (1985), *Images of Science*. (Chicago, IL: University of Chicago Press).

Collins, H.M. and T.J. Pinch. (1982), *Frames of Meaning*. (London: Routledge & Kegan Paul).

DeMey, Marc (1982), *The Cognitive Paradigm*. (Dordrecht, Holland: D. Reidel).

Edge, David and Michael Mulkay. (1976), *Astronomy Transformed: The Emergence of Radio Astronomy in Britain*. (New York: John Wiley & Sons).

Faust, David (1984), *The Limits of Scientific Reasoning*. (Minneapolis, MN: University of Minnesota Press).

Gamow, G. (1947), *One, Two, Three . . . Infinity*. (New York: The Viking Press).

Gardner, Howard (1985), *The Mind's New Science*. (New York: Basic Books).

Goodman, N. (1972), *Problems and Projects*. (Indianapolis, IN: The Bobbs-Merrill Co.).

Gruber, H.E., and P.H. Barrett. (1974), *Darwin on Man*. (New York: Dutton).

Gruber, H. and J.J. Voneche, eds. (1977), *The Essential Piaget*. (New York: Basic Books).

Habermas, Jurgen (1971), *Knowledge and Human Interests*. (Boston: Beacon Press).

Haugeland, John, ed. (1981), *Mind Design*. (Cambridge, MA: The MIT Press).

Hollis, M. and S. Lukes, eds. (1982), *Rationality and Relativism*. (Cambridge, MA: The MIT Press).

Hooker, C.A. (1978), 'An Evolutionary Naturalist Realist Doctrine of Perception and Secondary Qualities'. In *Minnesota Studies in the Philosophy of Science*, Vol. 9. C. Wade Savage, ed. (Minneapolis, MN: The University of Minnesota Press).

Johnson-Laird, P.N. (1983), *Mental Models*. (Cambridge, Mass.: Harvard University Press).

Kahneman, D., P. Slovic, and A. Tversky. (1982), *Judgment Under Uncertainty: Heuristics and Biases*. (Cambridge University Press).

Knorr-Cetina, Karin D. (1981), *The Manufacture of Knowledge*. (Oxford: Pergamon Press).

Knorr-Cetina, K.D. and Michael Mulkay, eds. (1983), *Science Observed*. (Hollywood, CA: Sage Publications).

Kornblith, H., ed. (1985), *Naturalizing Epistemology*. (Cambridge, MA: The MIT Press).

Kuhn, T.S. (1962), *The Structure of Scientific Revolutions*. (Chicago: University of Chicago Press, 2nd ed., 1970).

Kuhn, T.S. (1974), 'Second Thoughts on Paradigms'. In *The Structure of Scientific Theories*. F. Suppe, ed. (Urbana: University of Illinois Press, 459–82).

Lakatos, I. (1970), 'Falsification and the Methodology of Scientific Research Programmes'. In I. Lakatos and A. Musgrave, eds. *Criticism and the Growth of Knowledge*. (Cambridge: Cambridge University Press).

Latour, Bruno and Steve Woolgar. (1979), *Laboratory Life*. (Beverly Hills, CA: Sage Publications).

Laudan, L. (1977), *Progress and Its Problems*. (Berkeley: University of California Press).

Laudan, L. (1984), 'Explaining the Success of Science: Beyond Epistemic Realism and Relativism.' In *Science and Reality*. J. T. Cushing, C. F. Delaney, and Gary M. Gutting, eds. (Notre Dame, IN: University of Notre Dame Press).

Laudan, L. et al. (1986), 'Scientific Change: Philosophical Models and Historical Research'. *Synthese* 69, 141–223.

Leplin, J. ed. (1984), *Scientific Realism*. (Berkeley, CA: University of California Press).

Lumsden, C. and E.O. Wilson. (1981), *Genes, Mind, and Culture*. (Cambridge, MA: The Harvard University Press).

McCarthy, Thomas (1978), *The Critical Theory of Jurgen Habermas*. (Cambridge, MA: The M.I.T. Press).

Minski, M. (1975), 'A Framework for Representing Knowledge'. In P. H. Winston, ed., *The Psychology of Computer Vision*. (New York: McGraw-Hill) 211–277.

Neisser, U. (1976), *Cognition and Reality*. (New York: W. H. Freeman).

Newton-Smith, W.H. (1981), *The Rationality of Science*. (London: Routledge & Kegan Paul).

Nisbett, R. and L. Ross. (1980), *Human Inference: Strategies and Shortcomings of Social Judgment*. (Englewood Cliffs, NJ: Prentice-Hall).

Pickering, A. (1984), *Constructing Quarks*. (Chicago: The University of Chicago Press).

Plotkin, H.C. ed. (1982), Learning, Development, and Culture. (New York: John Wiley & Sons).

Pylyshyn, Z. (1984), *Computation and Cognition*. (Cambridge, MA: The MIT Press).

Putnam, H. (1982), 'Why Reason Can't be Naturalized'. *Synthese* 52, 3–23.

Rensch, B. (1971), *Biophilosophy*. (New York: Columbia University Press).

Rorty, R. (1979), *Philosophy and the Mirror of Nature*. (Princeton: Princeton University Press).

Rubinstein, Robert A., Charles D. Laughlin, Jr., and John McManus. (1984), *Science as a Cognitive Process*. (Philadelphia: University of Pennsylvania Press).

Ruse, M. (1986), *Taking Darwin Seriously* (Dordrecht-Holland: Reidel).

Shapin, Steven (1982), 'History of Science and its Sociological Reconstructions'. *History of Science* XX, 157–211.

Siegel, H. (1980), 'Justification, Discovery and the Naturalizing of Epistemology'. *Philosophy of Science* 47, 297–321.

Siegel, H. (1985), 'What is the Question Concerning the Rationality of Science?' *Philosophy of Science* 52, 517–537.

Simon, H.A. (1957), *Models of Man*. (New York: John Wiley & Sons).

Stroud, B. (1981), 'The Significance of Naturalized Epistemology'. In *Midwest Studies in Philosophy*. (Minneapolis, MN: University of Minnesota Press). 455–471.

Tolman, E.C. (1948), 'Cognitive Maps in Rats and Men'. *Psychological Review* 55, 189–208.

Toulmin, S. (1972), *Human Knowledge*. (Princeton: Princeton University Press).

Tweney, R.D., M.E. Doherty, and C.R. Mynatt, eds. (1981), *On Scientific Thinking*. (New York: Columbia University Press).

van Fraassen, B.C. (1980), *The Scientific Image*. (Oxford: Oxford University Press).

# A Cognitive-Historical Approach to Meaning in Scientific Theories

NANCY J. NERSESSIAN

*Center for Philosophy of Science, University of Pittsburgh*

The creation of concepts through which to comprehend, structure, and communicate about physical phenomena constitutes much of the scientific enterprise. Concepts play a central role in the construction and testing of the laws and principles of a theory. The introduction of new concepts and/or the alteration of existing ones is a crucial step in most changes of theory. And, in many scientific controversies what is at issue is disagreement over the interpretation of fundamental concepts. In short, articulating concepts is a central aspect of scientific research. Thus, our understanding of science is seriously deficient if we fail to examine the question of how scientific concepts emerge and are subsequently altered. Yet, such examinations have played little role in the philosophy of science. This is especially surprising in view of the fact that problems of conceptual change in science, in the form of the problems of 'meaning variance' and 'incommensurability', have dominated so much of post-positivistic philosophy of science. Most responses to these problems have centered on discussion of the alleged nature and necessities of language *per se;* the presumption being that the results of such analysis can simply be transferred to the scientific case. Thus, actual linguistic practices in science, and in particular the processes of concept formation and change, have gone largely unexamined. The result of this neglect has been that philosophical accounts of 'meaning' and 'meaning-change' for scientific theories and scientific practices concerning meaning continue to be at odds with one another.

In neglecting to add the dimension of 'discovery' to accounts of 'meaning' and 'meaning-change' in science, the prohibitions of logical positivism continue to influence contemporary philosophy. Within the framework erected by logical positivism, change, continuity, and comparison of meaning presented no problem. The observation language of science provides a 'theory-neutral' basis to which all theoretical terms can ultimately be reduced. Concept formation in science was characterized as continuous and cumulative, with the new or altered conceptions being logical extensions of previous notions. The major problem was that of specifying the nature of the reduction of the theoretical terms to the observational. Forming the background of this enterprise was the distinction – codified by Reichenbach – between the 'context of discovery' and the 'context of justification'. Only the latter

*Nancy J. Nersessian (editor), The Process of Science. ISBN 90-247-3425-8.*

was held to be the proper domain of the philosopher. Study of the actual creation of scientific concepts, for example, was deemed the province of the historian and the psychologist. Thus, when Hempel wrote his classic treatise addressing the important problem of how scientific concepts are formed, the object of study was not the the the actual formation of scientific concepts, but rather the 'rationally reconstructed' conceptual structure of science.[1]

With the realization that it is not possible to defend such black and white distinctions as formed the context within which the 'received view' conducted its analysis, the problem of meaning-change – the *bête noir* of post-positivistic philosophy of science – came to the fore.[2] If no sharp distinction can be made between the theoretical and the observational languages of a scientific theory, there can exist no 'neutral' point of comparison for the terms of a theory and its competitor or successor. This view was buttressed by the claims of Feyerabend and Kuhn that analysis of the history of science reveals 'incommensurability' of meaning to be a fact. Change of meaning is the result of a catastrophic 'revolution'; i.e., it takes place in a discontinuous manner and in such a way that the concepts of the new theory completely replace those of the previous. For those who conceded incommensurability as an insuperable problem, the problem of meaning-change has been supplanted by that of replacement: of one 'incommensurable paradigm' or 'hard core of a research program', etc., by another. For those who are still concerned to establish continuity despite incommensurability, the favored approach has been to provide for continuity of reference while admitting incommensurability of sense or intension. The most well-developed of these attempts is the 'causal theory' of Kripke and Putnam in which sameness of 'rigid designators' is what provides for continuity. However, this theory, formulated initially in response to problems in the philosophy of language, has serious difficulties when transferred to the domain of science.

Rather than recount the deficiencies of past and current accounts of 'meaning' and 'meaning-change' for scientific theories, my intention here is to argue for the appropriate method by which to obtain an adequate account (Part I) and to illustrate the method with two specific problems: how to provide an adequate representation for 'the meaning' of a scientific concept (Part III) and how to analyze the role of imagery and of analogy in concept formation in science (Part IV).

# I

The problems which need to be addressed by an account of 'meaning' and 'meaning-change' in science divide into to those that are 'representational' and those that are 'developmental' in character. Representational problems address such questions as how to specify the constituents of the meaning of a scientific concept and where to locate these concepts (i.e., in the heads of scientists? in the scientific community? in a Platonic realm?). Developmental problems are concerned with the processes through which concepts emerge and are subsequently altered. The specification and solution of both representational and developmental problems require a method which is 'cognitive' and 'historical'.

To begin with, the method is thoroughly historical. This is one instance in which the strong claim can be made that historical analysis is necessary for the philosophical problem. The problems (some of letters) or resolved merely by comparing the developed concepts of various theories. To take a well-worn example, we cannot simply compare Newtonian 'mass' to relativistic 'mass'; rather, what is called for is a detailed examination of how we got from one to the other. The accounts which have claimed to establish incommensurability as a fact of history are decidedly *un*historical. What is lacking from them are fine-structure analyses of the period of transition between theories. Where these have been provided, the 'problem of incommensurability of meaning' appears not all that profound. Indeed, what examination of the history of science shows, when we take into account that, for example, we did not go directly from Newtonian mechanics to the special theory of relativity, but passed at least through the theory of electrons, is that the real problem is one of accounting for continuous development which is not simply cumulative; i.e., of accounting for 'meaning variance' with commensurability.

There has been some tendency for post-positivistic philosophers of science to turn to case studies of particular episodes in the history of science to test philosophical theories. Many of these have functioned as counterexamples, and as such they have been of value to philosophy of science. Others have unfortunately been used as grist for philosophical mills. In such instances, as has been pointed out repeatedly by historians of science, the historical data have been distorted to fit the needs of a particular philosophical theory. A quite different approach is being argued for here. The history of science needs to be examined with a philosophical sensitivity to the problems at hand, but not with the goal of fitting the data to presupposed solutions. We should be open to the possibility that some of our problems are not real and have a willingness to abandon proposed solutions as well as problems. It is, of course, not possible to come to the history 'pure', but as far as possible we should be cognizant of our preconceptions and try not to adapt the data to these. Our proposals should grow out of the historical data and should not be imposed upon them. Rather than only 'the historian's task', the 'spelling out of how particular concepts developed with time' is the starting – and end – point of a philosophical analysis of meaning in scientific theories.[3]

Philosophers have handicapped themselves at the outset by the presumption that the study of how concepts are formed, changed, and function in scientific practice is irrelevant to philosophical analysis. This presumption was explicit in logical positivism with its division of 'contexts', but it is also implicit in the more contemporary predilection for 'science fiction' examples in place of 'science fact'. The prohibition against discovery analysis is, however, misconceived. While it is true that there is an important distinction to be made between how one arrives at an hypothesis and whether it is justified, this does not entail that 'discovery' is irrelevant to either that question or to all philosophical questions. It is essential to an account of 'meaning' and 'meaning-change' as they relate to science. Reichenbach was right, 'discovery' is the domain of history and psychology, but here is a point where they interact with philosophy.

This brings us to the second half of the methodological proposal. The presumption of the cognitive aspect of the method is that the cognitive mechanisms at work in the meaning-making dimension of science cannot be fundamentally different, i.e., different *in kind*, from those we employ in non-scientific and science-learning contexts. Thus, the problems encountered in characterizing the nature and processes of concept formation and change in science can and should be examined in light of pertinent results, interpretations, and debates in the cognitive sciences, in particular, in cognitive psychology and in artificial intelligence, where it interfaces with psychology. A *caveat* is needed though. Cognitive science has certainly not advanced to the stage where we could even consider wholesale importation of analyses of human cognitive mechanisms to the scientific case. Nor is it desirable to do so. Any adequate science of cognition must also take the data from the analysis of science into account in its formulation, and this has not been done to any significant extent.

As a methodological proposal, what is being advocated here is not all that radical. Neopositivist theories of meaning made use of behaviorist psychology and some of their critics made use of *Gestalt* – though neither used psychology in a systematic and serious way. However, we can fully expect that some of the conclusions from the employment of a 'cognitive – historical' method will be radical. To mention just one that extends beyond the issues at hand, an adequate conception of 'scientific progress' will have to include the characterization of the history of science in terms of 'cognitive development'.

In conclusion, to paraphrase Putnam,[4] traditional philosophical accounts of 'meaning' and 'meaning-change' for scientific theories have left out actual scientific practices and the human beings who invent them. To rectify this a methodological proposal has been made: to fully incorporate the dimension of discovery – the history of science and the science of cognition – into the philosophical analysis of the conceptual dimension of science. Employing a 'cognitive – historical' approach to meaning in scientific theories is a multidisciplinary enterprise in which all of the disciplines stand to benefit from the interaction.

## II

Most of my efforts thus far have gone into providing a detailed analysis of the formation of and changes in the major conceptual innovation of modern physics: the field conception of forces.[5] Through this analysis certain representational and developmental problems have been uncovered. In the following sections I will discuss some of these problems. However, some preliminary discussion of the historical case would be helpful.

The initial objectives were to locate the major shifts in the meaning of 'field' in the science of electromagnetism between Faraday and Einstein and to examine the reasoning leading to the initial formulation and subsequent alterations of that concept. The concept of electromagnetic field was selected for analysis because it is the fundamental innovation of modern physics. Its development was an essential step in the modern 'scientific revolution'. Without its formulation relativity theory and quantum mechanics would not have been possible. Additionally, the development of the modern electromagnetic field concept spans several 'research programs'. Finally, the field concept is still undergoing development in the search for a unified field conception of forces. Given its central position in modern physical theory, we can learn much about the conceptual dimension of science by examining its development and its role in the formulation of several theories.

The contributions of Faraday, Maxwell, Lorentz, and Einstein were the focus of the study because, after a more extensive analysis, it appears that the most significant changes in meaning were made by these four. Although embryonic field concepts were around before Faraday, the analysis begins with his conception because it is here that the notion that an understanding of processes in the region surrounding bodies and charges is required for the description of electric and magnetic actions first took hold. Let me summarize briefly the different field conceptions in this historical period. In Faraday's theory, 'electromagnetic field' (which he called 'physical lines of force') is primarily a qualitative concept, i.e., of unknown mathematical structure. The field is either a property of an aether quite unlike ordinary matter or a substance, with Faraday's preference being to consider the lines of force as substances and force as the only substance. Thus, in his most speculative conception, the field concept would be extended to comprise all the forces of nature. With Maxwell 'electromagnetic field' becomes a quantitative concept. The field conveys electric and magnetic forces and light according to a specific set of mathematical formulae (Maxwell's equations). It is a state of a mechanical aether, that is, one which obeys Newton's laws, and is not clearly differentiated from ordinary matter. The mechanical processes are unknown stresses and strain in the medium. With Lorentz' theory of electrons, the electromagnetic field is a state of a rigid (non-Newtonian) aether and is produced in the interaction of charged matter and the aether, which is now clearly distinguished from ordinary matter. This interaction is governed by Maxwell's equations and a new non-Newtonian force law (the Lorentz' force), which violates action and reaction. In Einstein's special theory of relativity the electromagnetic field has the same mathematical structure as in the theory of electrons, but it is no longer a state of a quasi-material substance. Rather, it is ontologically on a par with matter; it is an irreducible element of description.

I want to emphasize two fundamental conclusions from this analysis before going on to examine specific problems. First, the creation of a scientific concept takes place within frameworks of beliefs and in response to specific problems. These beliefs and problems are theoretical, experimental, methodological, and metaphysical in nature. We cannot divorce questions of meaning from questions about these. When we examine the reasoning leading from one phase of development to the next, we see that changes in meaning come about in response to changes in beliefs and problems. For example, Maxwell altered Faraday's field concept in response to his belief that the new conception must be incorporated into the Newtonian framework and the problems this engendered. These changing networks of beliefs and problems are in part communal and in part individual. They overlap in such a way as to provide continuity and diverge enough to create fundamental changes. If we are to understand 'meaning' and 'meaning-change' as they relate to science we must include these networks in our analysis.[6] Second, in each theory the meaning of 'electromagnetic field' differs significantly. Yet, each concept shares in part of the meaning of its predecessors – though it shares more with its immediate predecessor than with those more remote. Thus, while there is a 'meaning variance' there is a significant degree of commensurability.

## III

The most pressing representational problem is that of specifying what constitutes 'the meaning' of a scientific concept. A prerequisite to any characterization of how concepts form and change is a determination of what is to be included in the meaning of a concept. To allow for an account of 'meaning-change', the representation of a scientific concept must be both synchronic and diachronic. That is, the representation must stipulate what constitutes 'the meaning' of a concept at a point in history and as it changes over time, either intra- or intertheoretically.

This is a problem of central concern for historical method as well. A prime concern of historians of science is to avoid committing the 'Whiggish fallacy'; that is, attributing present-day notions to past science. In discussion of scientific concepts one must be particularly alert. Past concepts seem quite familiar but nevertheless may differ in significant respects from modern ones. Yet, historians of science possess no explicit metatheoretical account of 'the meaning' of a concept which will help to avoid this pitfall; and, as I have shown in my analysis of the controversy over when Faraday had his field concept, even the best historians fall into the trap because they have not articulated what it is to 'have' a concept.[7] To ask at what point did X have concept 'Y', assumes not only that we have criteria for 'Y' but also that we have criteria for the form of representation of 'the meaning' of a scientific concept in general.

The form of the synchronic representation must be determined first. The most widely assumed (at least tacitly) notion is that a concept is represented by a 'definition': by a set of conditions each of which is necessary and all of which are jointly sufficient to define it. With few exceptions, philosophers and historians of science, and, until quite recently, psychologists as well, have accepted this meta-theoretical notion uncritically. Yet, when we examine this notion, we see that it presents serious problems when applied to scientific concepts. In part, it contributes to the 'problem of incommensurability of meaning' between scientific theories; for if a different set of necessary and sufficient conditions are ascribed to 'Y' 'in theory T and 'Y'' in theory T', how can the 'Y's' be related? And, of course, there is the question of how to select the necessary and sufficient conditions which define a concept. Do we select them to match our modern concepts? If so, then we have to say that no one had a field concept before Einstein because no one used all the modern criteria before him. The alternative is to make Faraday's concept seem identical to Einstein's at least in what we take to be its 'essential' features (assuming we can determine these!), but this would seriously distort the historical data. If, on the other hand, we choose the conditions from an examination of each particular conception, we then have many different so-called field concepts throughout the history of science without being able to say how they are related. Use of the term 'field' alone is not sufficient to provide the connection.

All of this, of course, assumes that it is even possible to state a set of necessary and sufficient conditions which define a particular concept. Yet, as has been argued many times, it must be conceded that this is notoriously difficult, if not impossible, to do except, perhaps, in the case of concepts introduced by explicit definition – which happens not all that frequently in science. Central among the difficulties is that of how to distinguish between an 'essential' and an 'accidental' feature. That is, between one which all instances must have and one which most instances do have. To take a simple example, 'flies' is a feature of most birds, but it is not essential since some birds do not fly.

The 'definitional' notion of the representation of a concept does not do justice to the data of the electromagnetic field study. Although the details of a satisfactory representation have yet to be worked out, let me first present the analysis of these data as it now stands, and then discuss a promising avenue of resolution for this representational problem. First, we need to distinguish between the general concept 'electromagnetic field' (type) and instances of it (tokens), such as the electromagnetic field concept of the theory of electrons. (Note that we can carry the abstraction further here: 'electromagnetic field', 'gravitational field', etc., are instances of 'field'.) For each instance, i.e., synchronically, its meaning can be represented by a multicomponent 'vector'. Just what are the salient components of the vector remain to be determined. These may be different for different kinds of concepts, such as substantive concepts and dispositional concepts. For a quantitative substantive concept such as 'field', some components would be its mathematical structure, causal properties, ontological status, and reference. To represent the meaning diachronically we can expand the vector into an 'array' – what I have elsewhere called a 'meaning schema'[8] – which exhibits how the various components changed or did not change over time. In sum, the representation of tokens of scientific concepts would be the set of salient components of its 'vector' and that of the type, a set of family resemblances exhibited in its 'meaning schema'. The problems of determining the components, of what is a 'resemblance', and of how to weigh the relative importance of resemblances remain.

Questions about what form the representation of a concept takes have been the subject of much recent debate in cognitive science. This discussion has been sparked in part by the work of the psychologists Rosch and Mervis on the role of prototypes in categorization[9] and by renewed interest in the later work of Wittgenstein.[10] What has emerged from this discussion is that it is possible to represent concepts in such a way that while they are undefinable, they are complex – although there is far from a consensus on how to do this. Injecting the requirements of the representation of scientific concepts into the debate should benefit both enterprises.

The predominant conceptions are the 'prototype' view and the 'probabilistic' view.[11] In both cases, the representation of a concept is a set of weighted features. On the 'probabilistic' view, this set of features is fundamental, and a prototypical instance, i.e., one which maximizes the weighted score, is generated as a by-product. With the 'prototype' view, the prototypical instance is itself basic, and it determines the weight of the various features. In both cases, the features which represent the concept are those which have a substantial probability of occurring, rather than those deemed 'essential' or 'necessary'. The overlapping set of resemblances makes the concept into a unit and entitles us to call it the 'Y'. Additionally, the possibility is open that the representation can change over time. So, a concept can be represented both synchronically and diachronically, which scientific concepts require.

Of the two views, the 'probabilistic' seems more in accord with the scientific case. The set of features is fundamental, and a prototype would be generated from the 'meaning schema'. Whether this is so and how to select the salient features and assign weights require further analysis. The main point I wish to make here is that either view is better suited to our purpose than the 'definitional' view. They both offer the potential to represent development, continuity, and change in a way that the definitional notion cannot. We can, for instance, attribute a field concept to Faraday quite early without having to attribute either all the features of his mature concept or those of the modern conception. We can show that its specific features developed over time and that it has features quite unlike other field concepts; yet still maintain that it is connected with other field concepts, in particular, the modern one. In this way the 'problem of incommensurability of meaning' is overcome: The earlier and later forms of a concept bear a familial relationship to one another, and that is sufficient.

## IV

I want now to discuss some developmental problems. By what processes are scientific concepts formed and changed? Contrary to those who have claimed that new concepts arise by some inexplicable 'creative leap' that defies rational analysis, when we examine the reasoning leading to the emergence and alteration of concepts, it is possible to discern methods of concept formation which are subject to rational analysis. Study of these methods shows that the dynamics of change are not those of abrupt, cataclysmic revolutions. However, such analysis does not yield Baconian 'sausage-making machines' for concepts either. Rather, what we have are certain recurrent strategies for the articulation of scientific concepts. The employment of such strategies does not, of course, take anything away from the 'creative' nature of concept formation, since development, refinement, and application of these is certainly a creative enterprise. Study of the various strategies of concept formation will also provide substance to the claims of psychologists that concepts are formed by 'combination', 'differentiation', and 'shifts' from existing concepts. What is missing, by-and-large, from these claims is an account of how or by what processes 'combined', etc. I will begin with a report of my analysis of the historical data and then present a more general analysis, making use of relevant work in cognitive science.

Two strategies will be discussed here. It is clear from the Faraday – Maxwell portion of my study that the use of imagery, which I limit here to mean 'pictorial representations',[12] and of analogy played a significant role in the construction of their respective 'field' concepts. Two images figured centrally in this case: lines of force and interlocking curves (see Figures 1 and 2). In providing a symbolic representation of the actual mode of transmission of electric and magnetic actions, features of the images were incorporated into the two field concepts. At the outset, the image of the lines of force provided Faraday with a means of selecting terminology with which to discuss and reason about the phenomena and to communicate his interpretation of them to others. In discussing the actions he used such terms as 'moving out', 'expanding', 'collapsing', 'bending', 'straining', 'vibrating', 'being cut', and 'turning

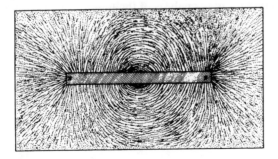

*Fig. 1.* Lines of force surrounding a bar magnet.

corners'. However, more is revealed in his selection of terminology than would be in a more metaphorical use of language. Faraday's field conception involves the notion that electric and magnetic actions actually take place along or through curved lines of force. The curvature of the lines was an essential component of his argument that these actions are not actions at a distance, and he attempted many times to demonstrate the curvature experimentally.

Much of Faraday's efforts went into trying to formulate a coherent conception of how *motion* of the lines could account for all the forces of nature. The relationship between static and dynamic electricity might be found in the expansion and collapse of the lines of force; while magnetism might consist in a 'vibration' of the lines and might be connected with a lateral repulsion between them. Electrostatic induction might take place through the action of contiguous particles along curved lines of force, and magnetic induction, by means of 'conduction' of the lines with varying degrees of ease through different media. Light and gravitation could be the result of a 'shaking' or 'vibrating' of the lines. Even matter itself might be nothing more than point centers of converging lines of force.

Further, there is a significant connection between the image of the lines of force and the only quantitative measure introduced by Faraday. The image makes the lines appear discrete, while they actually spiral indefinitely in a closed volume. Faraday's relationship is between the number of lines cut and the intensity of the induced force. This relationship is incorrect because 'number of' lines is an integer, while 'field intensity' is a continuous function. At the very least, he should have said 'number of cuttings', since the same line can be cut more than once. Finally, the influence of the image is apparent in the analogies he selected to describe the type of action he meant. He chose other 'line-like' phenomena such as rays of light and heat, rings in water, and conduction through wire.

Near the end of his research, Faraday introduced the representation that was to play a central role in Maxwell's reasoning. He represented the reciprocal nature of the actions of electricity and magnetism by interlocking curves at right angles to one another (see Figure 2).[13] The image of what Maxwell called *'mutually embracing curves'*[14] (Maxwell's italics) of electricity and magnetism figured strongly in his early attempts to give a quantitative analysis of the lines of force as representative of the

*Fig. 2.* Faraday's pictorial representation of the interconnectedness of electric currents and magnetic force.

'intensity and direction of electric and magentic forces at a point in space, and vestiges remain in his dynamical account.[15] Faraday's own discussion makes clear the relationship depicted between electricity and magnetism:

> Now these effects [repulsions and attractions] are not merely *contrasts* continued through two or more different relations, but they are contrasts which *coincide* when the position of the two axes of power at right angles to each other are considered. The tendency to *elongate* in the electric current, and the tendency to *lateral* separation of the magnetic lines of force which surround that current, are both tendencies in the same direction, though they seem like contrasts, when the two axes are considered out of their relation of mutual position; and this, with other considerations to be immediately referred to, probably points to the intimate physical relation, and it may be, to the oneness of condition of that which is apparently two powers or forms of power, electric and magnetic. (Faraday's italics).[16]

Here we have the basis for Maxwell's development of a reciprocal dynamics of electromagentism: A lateral repulsion between the magnetic lines of force has the same effects as a longitudinal contraction between current lines, and *vice versa* (see Figure 3). Without going into detail, the influence of the image on the mathematical representation developed by Maxwell can be seen in his complicated use of two fields each – one for 'intensity', a longitudinal measure of power, and one for 'quantity', a lateral measure of power – for the electric force and for the magnetic force. And, in his later work, the image provided a means of representing the notion that electromagnetic actions are propagated through the aether and that light itself is just such an action. We begin with an electric field and a magnetic field, and then

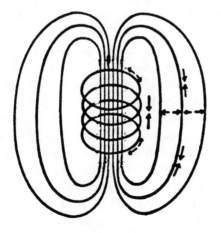

*Fig. 3.* Schematic representation of the reciprocal relationship between magnetic lines of force and electric current lines. Reproduced from Wise (1979).

summations of the quantities and intensities associated with these are propagated through the aether (expand Figure 2 into a 'chain' in the direction of propagation). As Norton Wise has remarked in his article on Maxwell's use of this image, the 'mutual embrace now became productive of offspring'.[17]

Use of analogy played the crucial role in the transition from Faraday's qualitative concept to the quantitative concept of Maxwell. Here analogy performed what I call 'assimilating' and 'articulating' functions. That is, the analogies provided temporary physical and mathematical meaning to the general conception of the continuous transmission of electric and magnetic forces through a mechanical medium by both assimilating the assumed relations to familiar ones and providing a basis from which to articulate the new meaning. A new concept was thus articulated by mapping the familiar relationships onto the new domain and fitting them to the requirements of the phenomena under investigation. What the new concept incorporates, and thus shares with the domain of the analogue, are certain abstractions expressed in the relationships. Maxwell called his own use of this method 'physical analogy'.

The method of 'physical analogy' is the working out of a possible mathematical representation of a set of physical phenomena on the basis of a partial isomorphism between the laws of a known set of phenomena and the relationships assumed to hold for a set of phenomena to be investigated. It is a 'method of investigation, which allows the mind at every step to lay hold of a clear physical conception; without being committed to any theory founded on the physical science from which the conception is borrowed.'[18] That is, it supplies a mathematical formalism plus a concrete physical image representing the assumed relationships to apply in the investigation, without requiring that a physical hypothesis be made. Employment of this method in the construction of a quantitative field concept enabled Maxwell to exploit the powerful representational capabilities of continuum mechanics.

It is not possible to go into the details of Maxwell's analogies here, since they are too complicated to allow a brief discussion. I have examined some of them in detail in my book.[19] The outlines of his method are as follows. In his first paper,[20] Maxwell tried to find a mathematical formulation of the notion that the lines of force represent the intensity and direction of the force at a point. He used an analogy between the intensity and direction of a line of force at a point and the flow of an incompressible fluid through a fine tube of variable section. In this way, Faraday's relationship between the number of lines cut and the induced force is replaced by a continuous measure. The analogy of the second paper,[21] in which he assumed that the field has physical existence, provided a means of expressing certain potential stresses and strains in a mechanical electromagnetic medium in terms of well-formulated relationships between known mechanical phenomena. The pictorial representation of this analogy is given in Figure 4.

The second analogy was designed to take the following into account: (1) a tension along the lines of force, (2) a lateral repulsion between them, (3) the occurrence of electric action at right angles to magnetic, and (4) the rotation of the plane of

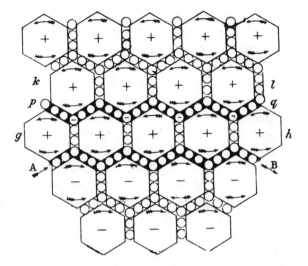

*Fig. 4.* Maxwell's pictorial representation of the assumed stresses and strains in the aether.

polarized light by magnetic action. Maxwell maintained that a mechanical analogy consistent with these four constraints is that of a fluid medium composed of elastic vorticies under stress. The problem of how to construe the relationship between electric currents and magnetism required a modification of the analogy. Since contiguous parts of the vorticies must be going in opposite directions, mechanical consistency requires the introduction of 'idle wheels' to keep the rotation going. Maxwell conceived of these 'idle wheels' as small, spherical particles, surrounding the vortices and revolving in the direction opposite to their motion, without slipping or touching each other. There is a tangential pressure between them and the vorticies. The relationships expressed here are those between a current (spherical balls) and magnetism (vorticies). Finally, his analysis of electrostatic induction required that the analogy once again be modified to express the necessary relationships. Now the entire medium was made elastic.

By using the method of 'physical analogy' Maxwell transformed the qualitative electromagnetic field concept of Faraday into a powerful quantitative concept fundamentally different from it. There were still some important unclarities such as just what are the properties of this medium, the aether, in which the field is produced; and even Maxwell considered his conception to be incomplete. With hindsight we can see that the vaguenesses in this conception would be clarified in a quite unexpected way. Maxwell made the assumption that the field is a state of an aether that is a mechanical system, i.e., obeys Newton's laws. However, the common abstraction in the analogy between the assumed stresses and strains in the medium and the familiar relationships between the stresses and strains as formulated in continuum mechanics is that of a more general dynamical system – what Maxwell called a 'connected system'. And, as was worked out in the period between Maxwell and Einstein, Newtonian mechanics is just one instance of such a system; relativistic mechanics is another. Thus, the relationships expressed in the laws of electromagnetism are those of a dynamical system, but one which is non-Newtonian, i.e., contrary to Maxwell's assumption.

The issues involved in analyzing the role of imagery and analogy in concept formation in science are ones on which recent work in cognitive psychology can be brought to bear. Analogies, in their assimilating and articulating functions have much in common with the 'structure-mapping theory' of analogies proposed by Dedre Gentner and collaborators.[22] This theory was designed to account for empirical studies of how people use analogies in learning new scientific concepts. Although this cognitive activity and that of creating the concepts are not identical, they are related, significantly, to one another. Comparison of the use of imagery and analogy in the articulation of scientific concepts with the claims of the structure-mapping theory helps to show how. Although the structure-mapping theory as presently formulated is not wholly adequate to the needs of the 'field' case, when combined with these data it offers promise of a deeper understanding of the roles of imagery and analogy in concept formation in science – both for scientists trying to construct a new conceptualization and for students attempting to understand extant scientific concepts.

The structure-mapping theory of analogies focuses on the insights that (1) scientific analogies are systems of connected knowledge and that (2) in its meaning-giving aspect, evaluation of an analogy is in terms of its explanatory power and not its validity as an argument. Briefly, the theory assumes that the predicates of the base domain are brought across to the target domain as identical matches and not similarities. It assumes that 'similarity' can be construed as an identity between some number of component predicates. The difference between an analogy (a 'non-literal similarity') and what is called a 'literal similarity' is that the predicates that are mapped in an analogy are mostly relationships and not attributes, whereas with a literal similarity many more attributes are brought across. A simple example of a literal similarity is one between our solar system and one in another galaxy; of an analogy, one between a solar system and an atom.

I intend to construct a detailed analysis of 'physical analogies' considered as 'structure maps' elsewhere. Here I want only to provide an indication of the potential fruitfulness of this theory to understanding the roles of imagery and analogy in scientific concept formation. To begin with, the analogies used by Maxwell are systems of connected knowledge. They are also not used as arguments, but as heuristic devices for exploring certain representational possibilities. Thus, they conform to the type of analogy examined by the structure-mapping theory. In the process of articulation of the field concept, certain features of the domain of the analogue were incorporated into it. These were primarily relationships. For example, a current does not have the attributes of the little spherical balls used to represent it; rather, the dynamical relationships between the balls and the vorticies are those assumed to hold between an electrical current and magnetism. One way to interpret Maxwell's repeated claim that no physical hypothesis is being made is as a claim that attributes are not being mapped from the domain of the analogue. Also, the structure-mapping theory holds that analogies mapping systems of relations (2nd-order or higher relations), such as causal relations, have higher explanatory value. This is borne out in the Maxwell case, where the analogy of the second paper went further in the articulation of the field concept by assuming identities in causal relationships between the two domains.

The continuum proposed by the structure-mapping theory between a literal similarity (mapping mostly properties) and an analogy (mapping mostly relations) offers the possibility of treating the roles of imagery and analogy on a continuum as well. Many of the attributes of the image of the lines of force are mapped into Faraday's field concept. With the images used to represent Maxwell's physical analogies, it is the relationships represented by the image, rather than the attributes, which get mapped. So, the function of the imagery has progressed from concrete to abstract. Images can be considered to be 'literal similarities' when they function concretely, and 'analogies' when they function abstractly. In its capacity as a literal similarity, the specific image is important, since many of the attributes it represents are incorporated into the concept. In its capacity as an analogy, it should be possible to employ any image which represents the relationships, since only the relationships are incorporated into the concept. Finally, changes in the function of the imagery employed might mark the transition from a qualitative formulation of a concept to a quantitative.

There are important issues pertaining to concept formation in science which the structure-mapping theory does not address. Among these are how an analogy can be altered to fit the needs of the domain under investigation, i.e., there is a dynamic aspect to this use of analogy which must be accounted for, as well as how the selection of a particular analogy takes place, which calls for an analysis of the role of beliefs and problems in the selection. In seeking answers to these and other questions raised in the examination of concept formation in science further application of the 'cognitive – historical' method is required.

## Acknowledgements

Drafts of this paper were read at the Center for Philosophy of Science, University of Pittsburgh, the University of Maryland, Cornell University, Rice University, and the APA/SWIP session honoring Marjorie Grene. I appreciate the comments made on those occasions and the comments made by Ryan Tweney on Part IV. I want, especially, to thank William Berkson for pressing me to articulate the problems of my research program.

176

## Notes

1. Hempel (1952).
2. For an analysis of the relevant distinctions and the implications of their collapse, see Nersessian (1984), pp. 5–23.
3. Enfield (1985), p. 641. Contrary to his claim that the 'philosophical reward' of such analysis 'seems slight', I contend that when the philosopher becomes an historian as well, the rewards will be great.
4. Putnam (1975).
5. Nersessian (1984), pp. 33–142.
6. Thus, contrary to Enfield's claim that '[a]stonishingly, Nersessian seems not to have noticed the resemblance between the network view and her own view that meanings involve networks of problems and beliefs' (Enfield (1985), p. 642). I, of course, recognize that my view is a network view – I in fact used the word 'network'. The analysis of the development of the field concept shows how this kind of network view escapes the incommensurability problems inherent in the Duhem – Quine and Kuhn – Feyerabend versions. What we end up with is a set of reason-related criteria for what a 'field' is. The problem then is to construct a representation for 'field' which encompasses these criteria. Referential stability is not a problem because continuity of reference is built into the synchronic and diachronic representation of the concept.
7. Nersessian (1985).
8. Nersessian (1984), pp. 153–160. Since developing this notion it has been pointed out to me that my notions of 'vector' and 'meaning schema' have much in common with the notions of 'schemata', 'frames' and 'slots' developed in cognitive science. That this is so underscores the fruitfulness of the 'cognitive – historical' approach. This representation, which grew out of the historical data, can be developed further by notions formed in cognitive science.
9. See, e.g., Rosch and Mervis (1975).
10. Wittgenstein (1953), especially section 65–88.
11. For a review of these positions, see Smith and Medin (1981).
12. For an extensive study of the role of visual thinking in the creation of the concepts of 20th century physics, see Miller (1984). Miller uses the term 'imagery' in a wider sense than I do here.
13. Faraday (1852), §3265. For a discussion of Faraday's earlier imagery, see Tweney (1985).
14. Maxwell (1856), p. 194n. Here Maxwell refers to Faraday's introduction of the image in his 1852 paper.
15. For an excellent analysis of the effects of the image on Maxwell's quantitative analysis, see Wise (1979).
16. Faraday (1852), §3268.
17. Wise (1979), p. 1316.
18. Maxwell (1856), pp. 155–156.
19. Nersessian (1984), pp. 69–94, 148–150.
20. Maxwell (1856).
21. Maxwell (1862).
22. The theory is developed over several papers, and with other researchers. Gentner (1983) gives the most succinct presentation.

## Bibliography

Enfield, P. (1985), *Review of Faraday to Einstein: Constructing Meaning in Scientific Theories*, in *Philosophy of Science* 52: 641–642.
Faraday, M. (1852), 'On the physical character of the lines of magnetic force', in *Experimental Researches*

*in Electricity* (New York: Dover, 1965), vol. 3, pp. 407–437.

Gentner, D. (1983), 'Structure-mapping: A theoretical framework for analogy', *Cognitive Science* 7: 155–170.

Hempel, C.G. (1952), *Fundamentals of Concept Formation in Empirical Science,* vol. 2, no. 7, *International Encyclopedia of Unified Science* (Chicago: University of Chicago Press).

Maxwell, J.C. (1856), 'On Faraday's lines of force', in *The Scientific Papers of James Clerk Maxwell,* W. D. Niven, ed., (Cambridge: Cambridge University Press), vol. 1, pp. 30–74.

Maxwell, J.C. (1862), 'On physical lines of force', in *Scientific Papers,* vol. 1, pp. 155–229.

Miller, A. (1984), *Imagery in Scientific Thought: Creating 20th Century Physics* (Boston: Birkhäuser).

Nersessian, N.J. (1985), 'Faraday's field concept,' in *Faraday Rediscovered: Essays on the Life and Work of Michael Faraday, 1791–1867,* D. Gooding and F. James, eds., (London: Macmillan), pp. 175–187.

Nersessian, N.J. (1984), *Faraday to Einstein: Constructing Meaning in Scientific Theories* (Dordrecht: Martinus Nijhoff).

Putnam, H. (1975), 'The meaning of 'meaning',' in *Mind, Language, and Reality* (Cambridge: Cambridge University Press), pp. 215–271.

Rosch, E. and Mervis, C.B. (1975), 'Family resemblance studies in the internal structure of categories', *Cognitive Psychology* 7: 573–605.

Smith, E.E. and Medin, D.L. (1981), *Concepts and Categories* (Cambridge, MA: Harvard University Press).

Tweney, R. (1985), 'Imagery and memory in Michael Faraday's scientific thought', lecture delivered at the Boston Colloquium for the Philosophy of Science.

Wise, N. (1979), 'The mutual embrace of electricity and magnetism', *Science* 203: 1313–1318.

Wittgenstein, L. (1968), *Philosophical Investigations,* 3rd. edition, G.E.M. Anscombe, trans. (New York: Macmillan).

# Naturalizing Observation[1]

HAROLD I. BROWN
*Northern Illinois University, Dekalb, Illinois*

One of the major developments in recent philosophy is the movement towards a naturalized epistemology.[2] Although there is the usual disagreement in detail among proponents of this movement, one proposition is central: that human knowledge is a natural phenomenon. This, in turn, suggests an evolutionary epistemology based on the thesis that the human species appeared in this world without any privileged access to its nature, and without any privileged access to the best methodology for learning about the world. Through time we have learned much on both fronts, and the development of science in particular has increased both our knowledge of the world we live in, and our knowledge of how to learn about that world. Still, in both cases, our knowledge remains tentative and uncertain, and we have every reason to assume that continued research will continue to bring surprises. An immediate consequence of this is that we should not expect a higher degree or certainty in epistemology than in science, and one of the epistemologist's major concerns is to examine the best available examples of human knowledge as a basis for reassessing our current understanding of the way in which the knowledge enterprise proceeds. In this paper I want to consider scientific observation from a naturalistic point of view.

Much of the power of modern science derives from the use of instruments that are interposed into the causal chain between our senses and the world, and that permit us to observe items[3] that we cannot detect with our unaided senses. Some of these instruments, such as the telescope, improve our ability to observe items that we can, to some degree, observe with our senses alone, while others – e.g., the magnetic compass, the electron microscope, or the Geiger counter – permit us to observe items that are inaccessible to our unaided senses. Not all instruments work this way; a meter stick or an astrolabe, for example, does not introduce an additional causal link between our senses and the item under observation, and in this paper I will only be concerned with instruments of the former sort, i.e., those that do enter into the causal chain between observer and observed.

From the point of view of a naturalized epistemology our senses are natural objects that function in accordance with natural laws, and our instruments are extensions of these senses that permit us to observe items that we cannot observe

*Nancy J. Nersessian (editor), The Process of Science. ISBN 90-247-3425-8.*

with our senses alone. While I cannot attempt a full defense of this claim in a single paper, I will present the major reasons for adopting it in two stages: In part I, I will develop this approach by considering seven claims about the relation between our senses, our instruments, and the physical world. In part II, I will briefly consider three cases from the history of science in the light of this approach.

## I. Observation from a naturalistic point of view

1. *There are no a priori grounds for maintaining that the set of items that exists in the physical world is coextensive with those we can detect with our unaided senses.* It has taken a long time for the full significance of this claim to become clear. Common sense, along with much early philosophy and science, assumes that our senses functions as 'windows' on the world which simply show us what is there. This is not an unreasonable position in the absence of strong evidence to the contrary, but science has been providing such evidence at least since the seventeenth century. The earliest philosophical response to this evidence was the distinction between primary and secondary qualities, a distinction built on the suggestion that some of the items we sense – the secondary qualities – do not copy items in the physical world, and thus that the set of items in the physical world is *smaller* than the set of items we can sense. It is only more recently that the development of science has pressed on us the recognition that the items we perceive are, at best, only a small subset of the items that fill the world around us.

2. *Our senses are physical objects which operate in accordance with the same physical laws that govern the behavior of other physical objects – including the instruments that scientists design and deploy.* An immediate consequence is that we should not expect our senses to be somehow free of the kinds of defects and sources of error that occur in our instruments. To a degree, this is a familiar point. We recognize that the lenses of our eyes operate in accordance with the same laws of refraction as other lenses, and can, for example, fail to have the correct focal length for a particular job. And we have long known how to correct this particular defect by the application of supplementary lenses. We have also tended to assume that the 'normal' eye is a completely reliable[4] source of information about the objects it detects, but the earliest systematic use of a modern observing instrument – Galileo's study of the heavens with his telescope – provided evidence that, even within its 'normal' range, the eye can give misleading information. Let me illustrate the point by considering a problem that Galileo faced, and his proposed solution to this problem.[5]

Once the telescope was turned on the heavens, it was quickly noted that distant objects such as the fixed stars do not appear to be magnified to the same degree as nearer objects, and this was pointed to by some as evidence of the unreliability of the telescope in celestial observations. (Others suggested that this could be used as a means of determining the distance to the object in question.) Galileo responded by arguing that a given telescope does in fact enlarge all objects to the same degree, irrespective of their distance from the observer, and that the apparent differential enlargement is an illusion caused by a *defect in the eye* that results in a distorted image when we examine the star with the unaided eye. On Galileo's diagnosis, when we look at a bright object that has a small retinal image, the eye generates a fringe of light around the retinal image, and this causes the object we are seeing to look larger than it should. This occurs in naked eye observation of a star, and as a result, the star seems to be enlarged by the telescope less than other objects, although the star's retinal image is enlarged to the same degree as the image of any other object. Telescopic images are not distorted by this misleading fringe because the enlarged image provided by the telescope fills the retina, leaving no place where the fringe can be generated. Thus the value of the telescope goes beyond its ability to enlarge the images of distant objects – in addition, it compensates for a defect in the eye, and the telescopic image is thus a more reliable source of information about the star than the image provided by the unaided eye.[6]

3. A second consequence of point 2 is that *in all cases, including those in which we make use only of our native senses, what we observe is the result of a causal process involving our senses.* In the typical case, that in which we are observing an object that exists independently of us, what we actually sense is thus a function of *both* the properties of the item under observation *and* the properties of our senses. Once this has been recognized, it follows that we can never simply *assume* that the items we sense share any features of the items that initiated the causal process. This may occur; there are complex causal processes in which the final outcome of the process does indeed have properties that are close copies of properties of one of the items that entered into the causal chain – e.g., sound waves produced by a phonograph recording. But this need not occur, typically it will not occur, and we can never assume without independent evidence that it has (or has not) occurred. Yet even when the output of a causal chain does not share any properties of the items that entered into that chain, the output can still carry information about those items – and all observation consists of extracting information about items in a causal chain from a study of the last link in that chain. The introduction of instruments into such a chain may increase its length, and alter the properties of the output, but it does not involve the creation of a causal process where no such process had previously existed.

The above remarks suggest a straighforward response to a common argument on behalf of perceptual realism, i.e. that the fact of our survival shows that the information provided by our unaided senses must be a correct description of at least some items in our environment. Rather, all that is required to account for survival is that our senses provide useful information about the objects around us. Our ability to make use of this information no more requires that we perceive copies of items in our environment than does a pilot's ability to fly around a storm on the basis of a radar image require that the storm 'look like' the radar image. Nor need the pilot actually know what the storm is like in order to learn to avoid it. She might think that the storm is a set of shakings and tremblings sitting out in space, shakings and tremblings that she can directly feel if she is foolish enough to move into the portion of space that the radar image indicates they occupy.

Note also that, under appropriate circumstance, any item in the causal chain can become the item under observation. For example, the kind of causal chain that is used to study a star through an optical telescope can be used, on another occasion, to test the optics of a telescope – much as the visual procedures that we easily apply to find our way around the world can, in an optometrist's office, be used to gain information about the optics of our eyes.

4. A direct corollary of the last two points is that *we should not expect our senses to be capable of interacting with every item in our environment.* To a degree we are all aware of this, since we recognize that, for example, our eyes and ears respond to different feature of the world around us. But we should also recognize that it would be gratuitous to assume that our complete set of senses provides detectors for every item that exists. Our eyes only respond to a small piece of the electromagnetic spectrum, our ears are sensitive to only a small selection of the compression waves in our environment, and we have no built in detectors for neutrinos, viruses, radioactivity, or the toxicity of buried wastes. Indeed, it is reasonable to assume that the senses we do have evolved in response to the needs for survival in a particular environment, that if the environment had been significantly different we might have evolved different senses, and that if the environment changes dramatically, our unaided senses may no longer be sufficient to insure continued survival.

5. *The introduction of a complex causal chain can increase the reliability of an observation.* It is tempting to assume that this is not the case, i.e., that observations made with complex instrumental chains are *ipso facto* less reliable than those made with the unaided senses, and that we reach the limit of reliability when we make use of our senses alone. One traditional line of argument for this claim is based on the thesis that bare sense perception is infallible since it involves no use of tentative theories or learned conceptual structures and skills, and that the possibility of error enters in once these elements are invoked. Recent discussions of the 'theory dependence' of observation have, I believe, shown that this view is no longer supportable. However reliable our native sensory equipment may be, none of its outputs are relevant to epistemic concerns until they have been classified and understood in terms of some accepted body of knowledge. Thus we cannot divide off unaided sensory observation from instrumental observation on the basis of what is fallible and what is infallible. I will not pursue this line of argument any further here.

More to the point, it could be argued that even though observation with unaided senses involves the application of knowledge to the output of a causal chain, the introduction of instruments lengthens the causal chain and requires richer bodies of knowledge. Both of these factors open up additional ways in which we can go wrong, and thus greater possibilities for error. Moreover, all observations do pass through our senses, and any sources of error introduced by our instruments will thus be added onto those already involved in the use of our senses.

While this argument is seductive, it loses much of its force when applied to actual scientific cases. Consider again the example of Galileo and the telescope discussed above, a case which shows quite clearly that an increase in complexity of the causal chain can result in an increase in reliability. Contemporary astronomers often prefer to photograph the output from a telescope, and study photographs, instead of looking through the telescope. This added complexity *increases* reliability in many ways, not the least of which being that photography permits astronomers to deal with a stable, public object that can be examined and reexamined, instead of a fleeting, private visual image. This increase in reliability is really not surprising. We construct causal chains in order to achieve many different kinds of goals, and there are many cases in which a goal can be better achieved by a more complex chain than by one that involves fewer links. Moreover, there are many cases in which the argument in question is wholly beside the point. Without the aid of instruments we simply cannot observe such things as neutrinos, radioactivity, and the portion of a star's energy output that is outside the optical range. In such cases there is not much point to the suggestion that we lose reliability when we rely on instruments. To be sure, some of these observations are more reliable than others, but this is also true of observations made with our unaided senses, and the key point here is that it is a mistake to think that an increase in complexity of instrumentation or of background knowledge automatically entails a reduction in reliability.

6. *The causal process involved in an observation may be spread out over a substantial period of time.* Consider, for example, a standard sort of astronomical observation, in which a photographic plate receives the output from the telescope, while a tracking mechanism that compensates for the motion of the earth keeps the telescope pointed at the same portion of the sky for an extended period of time. One reason for doing this is to observe dim stars, i.e., stars that we cannot see, even when looking through the telescope, because their light output is too weak to trigger a visual response in us. By keeping the telescope pointed toward the relevant portion of the sky for a sufficient period of time, enough photons will be gathered to leave an inspectable image when the plate is developed. Other observational procedures take even longer, and there is no reason for imposing a time limit on those procedures that we are to count as 'observational'. If we find that this violates everyday linguistic sensitivities, or traditional conceptions of observation, we must conclude that those everyday and traditional notions are not adequate for the analysis of contemporary science. I will return to this point at the end of the paper.

7. *While it is true that our senses play a special role in observation, since any information derived from our instruments must pass through our senses, this is strictly a pragmatic matter, with no particular epistemological significance.* To see the significance of this point consider the possibility that, through genetic manipulation or microsurgery, we may find that we can dispense with some of the instruments that

we now require, and we may become able to sense 'directly' gamma radiation, or neutrinos, or a wider range of colors and pitches, or the direction in which the ambient light is polarized. If this were to occur, there would be important changes in the procedures by which we gather information, but the 'scope and limits of human knowledge' might well remain unchanged. Similarly, we may encounter aliens with different sense organs than ours, so that they need instruments where we do not, and *vice versa*. We would expect that the early history of their science would be different from the early history of ours, but once they have reached a sufficient level of instrumental sophistication, there is no principle of epistemology that prevents them from having arrived at the same views as we have for a wide variety of domains, independently of whether one group or the other can sense in that domain.[7]

Still, for the present, we are restricted to designing instruments that act as transducers between items we can detect with our senses, and other items that our senses cannot detect. Yet once we have mastered the use of such an instrument, we can use it to study items in our environment with as much confidence as we have in our evolved senses – indeed, with more confidence than we should have had in our senses when we understood less about how they operate. Some examples will illustrate these points.

Consider the magnetic compass, which literally lets us see the direction of the earth's magnetic field, even though our eyes do not respond to magnetism. How does it do this? It works because a magnetized needle is acted on by the local magnetic field in such a way that the needle lines up with the direction of the field. It happens that we can see the direction of the needle, and once we have recognized the correlation, we can look at the needle and read off the direction of the field. Our ability to do this is quite as reliable as, say, our ability to recognize a person or a tree when we see it from a distance.

In other cases the connection between the output of an instrument and the information we extract is not quite as direct as that between the direction of the magnetic field and the direction of the compass needle, but this should not mislead us. A Geiger counter, for example, responds to ionizing radiation and, by a somewhat more complex causal chain than that involved in the case of the compass needle, provides an auditory output from which we can learn of the presence and the strength of the local radiation. There is nothing 'auditory' about the radiation itself, but that is irrelevant. The Geiger counter provides a causal link between radiation that we cannot sense and compression waves that we can sense, and provides the trained user with reliable information about the radiation in her environment.

## II. Cases

My aim thus far has been to outline a naturalistic view of modern scientific observation, but the appropriateness and usefulness of this view can only be demonstrated by looking at actual scientific cases. Space limitations do not permit a detailed analysis of any case, but I do want to consider briefly three examples from physical science in order to illustrate the relevance of this approach. The first is somewhat unusual, but will be useful because it involves a case in which a problem appeared in what had been considered a straightforward use of our senses. The second is the unexpected observational discovery of a new type of particle, the positron. The third is the first observation of a long sought particle, the neutrino.

## 1. The personal equation

In 1796 the Astronomer Royal Nevil Maskelyne dismissed his assistant Kinnebrook because the latter systematically observed the times of stellar transits somewhat later than did Maskelyne himself; naturally the Astronomer Royal had reason to doubt his assistant's competence. Such observations were made in the following manner:

> The accepted manner of observing stellar transits at that time was the 'eye and ear' method of Bradley. The field of the telescope was divided by parallel crosswires in the reticle. The observational problem consisted in noting, to one tenth of a second, the time at which a given star crossed a given wire. The observer looked at the clock, noted the time to a second, began counting seconds with the heard beats of the clock, watched the star cross the field of the telescope, noted and 'fixed in mind' its position at the beat of the clock just before it came to the critical wire, noted its position at the next beat after it crossed the wire, estimated the place of the wire between the two positions in tenths of the total distance between the positions, and added these tenths of a second to the time in seconds that he had counted for the beat before the wire was reached. It is obviously a complex judgment. Not only does it involve a coordination between the eye and the ear, but it requires a spatial judgment dependent upon a fixed position (the wire), an actual but instantaneous position of a moving object, and a remembered position no longer actual. Nevertheless, 'the excellent method of Bradley' was accepted and regarded as accurate to one or at least two tenths of a second. In the face of this belief, Kinnebrook's error of eight tenths of a second was a gross error and justified Maskelyne's conclusion that he had fallen 'into some irregular and confused method of his own' and his consequent dismissal.[8]

Several years later Bessel became aware of the incident and began to explore the possibility that there was a 'personal factor' which varied from observer to observer using this method, and that perhaps by comparing the results obtained by different observers one could, in effect, calibrate the observer and thereby compensate for that portion of the observed result that was due to the astronomer, rather than to the motion of the star. Bessel did discover persistent differences between the results obtained even by those who were recognized as the most skilled observers, and expressed these differences in what came to be knows as a 'personal equation'. For example, Bessel made detailed comparisons between his own observations and those of Argelander, and concluded that, 'The personal difference between us is represented by the equation $A - B = 1.233$ sec'.[9] This sort of research only provides a relative personal equation, giving the relation between the observations of two specific individuals, but later investigators, equipped with more precise methods for determining the actual time of transit, attempted to determine absolute personal equations for specific observers. These attempts to calibrate observers met with only limited success, since observers show a considerable variability from observation to observation, and the need to rely on the astronomer's perceptual judgements in this case was eliminated by the development of electronic and photographic instruments which permitted more precise measurements while introducing further steps into the causal chain between star and human perceiver. As a recent writer has noted in a discussion of contemporary methods for observing star transits, 'If these depended ... on a human measurer, they are not likely to be accurate enough for modern purposes'.[10]

The crucial point I want to underline with this example is that the human observer, in all her complexity, is as much an active participant in the observation process as is the telescope or any other instrument. Thus the aim of achieving reliable information through observation often requires that we compensate for peculiarities of humans in general, and of particular humans in particular situations – just as we must compensate for systematic errors or other peculiarities of an observing instrument. Moreover, such seductive ideas as the view that our unaided senses are especially reliable, or that the reliability of an observation is inversely proportional to the amount of background knowledge required, are simply mistaken. In observation, as in many other situations, we increase reliability by drawing on the widest range of the best available body of knowledge, rather than by attempting to minimize our use of background knowledge.

## 2. The positron

Next consider C. D. Anderson's early observations of positrons, particles whose existence had not been recognized at the time in question.[11] The observation resulted from Anderson's examination of photographs of tracks in a vertical Wilson cloud chamber that Anderson was using to study cosmic radiation. The chamber was in a magnetic field of known strength and direction, and was bisected by a lead plate 6 mm thick. I want to examine Anderson's interpretation of one of these photographs in some detail (see Figure 1).

The photograph shows two curved tracks, one on each side of the image of the lead plate, and we should note that in examining the photograph Anderson had no way of determining directly the direction in which a particle was moving when a track was generated. Indeed, this was one of the central points at issue, since it was clear from

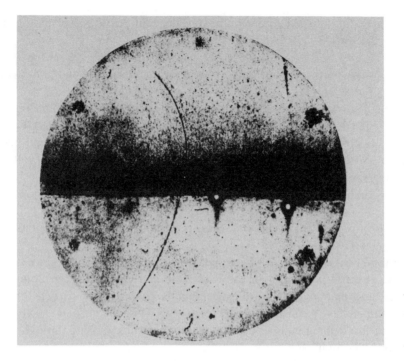

*Fig. 1.* From Anderson 1933, p. 492. Reprinted by permission of the American Physical Society.

accepted background knowledge, along with the known direction of the magnetic field, that a negatively charged particle must have moved from the top of the photograph towards the bottom, while a positively charged particle must have moved from bottom to top. Measurements of the length and curvature of the tracks, also in conjunction with accepted background knowledge, and a knowledge of the experimental apparatus, provided information on the particle's mass and charge, and eliminated the possibility that these tracks were caused by protons. Anderson writes:

> the track appearing on the upper half of the figure cannot possibly have a mass as large as that of a proton for as soon as the mass is fixed the energy is at once fixed by the curvature. The energy of a proton of that curvature comes out 300,000 volts, but a proton of that energy according to well established and universally accepted determinations [Anderson here refers to Rutherford, Chadwick and Ellis, *Radiations from Radioactive Substances*] has a total range of about 5 mm in air while that portion of the range actually visible in this case exceeds 5 cm without a noticeable change in curvature.[12]

In fact, it is clear that the mass of the particles involved is of the order of the mass of the electron, and this led Anderson to consider four possible explanations for the cause of the tracks:

(1) a positive particle of small mass penetrates the lead plate and loses about two thirds of its energy; or
(2) two particles are simultaneously ejected from the lead, in one direction a positive particle of small mass, and in the opposite direction an electron; or
(3) an electron of about 20,000,000 volts energy penetrates the lead plate and emerges with an energy of 60,000,000 volts, having gained 40,000,000 volts energy in traversing the lead; or
(4) the chance occurence of two independent electron tracks.[13]

Anderson now proceeds to eliminate those possibilities that are unacceptable. Let us follow his discussion.

The third possibility can be 'discarded as completely untenable'[14] since it requires that the electron pass through the lead plate and gain energy in doing so. All available information leads one to expect that the passage of the particle through the plate would involve a considerable *loss* of energy. This, of course, is exactly what we find if we are dealing with a positively charged particle passing through the plate from top to bottom.

If the tracks were not caused by a single electron, perhaps we have two distinct tracks, caused by two electrons, each moving up across the chamber. But then we must make sense of the high degree of precision with which the tracks line up:

This assumption was dismissed on a probability basis, since a sharp track of this order of curvature under the experimental conditions prevailing occurred in the chamber only once in some 500 exposures, and since there was practically no chance at all that two such tracks should line up in this way.[15]

Having eliminated the option that both tracks were caused by electrons, we turn to the second of the possibilities listed above, the possibility that the tracks were caused by two different particles that were simulataneously knocked out of a lead nucleus by a cosmic ray. This would account for the way the tracks line up, and would permit the moving particle that caused the lower track to be an electron. 'But in this case, the upward moving one would be a positive of small mass'[16] and would thus lead to the conclusion that such particles exist. And the first possibility, that both tracks were caused by a single positron moving through the lead, also leads to the conclusion that such particles exist. Thus the only plausible interpretations of the photograph require the introduction of a positively charged particle of very small mass, a particle whose existence had not previously been suspected.

We can see a similar reasoning process at work if we look, more briefly, at Anderson's handling of another photograph, this time one in which 'two tracks of opposite curvature appear below the lead'.[17] Again we have four possible accounts:

(1) a positive particle of small mass and an electron emerging from the same point in the lead; or
(2) a positive particle of small mass strikes the lead and rebounds with a loss in energy; or
(3) an electron of about 20,000,000 volts energy strikes the lead and rebounds with 30,000,000 volts energy; or
(4) the chance occurrence of two independent electron tracks.[18]

Here too, considerations of the curvature and length of tracks require that we are dealing with particles of very small mass; the first two possibilities involve the existence of a positron, as well as such physically reasonable assumptions as that a particle loses energy on a rebound. The third and fourth possibilities are attempts to get along with just electrons, and involve assuming that a particle might gain energy in rebounding, or that two independent tracks would line up in an improbably precise manner. Both of these assumptions can safely be rejected, and the reasonableness of attributing at least some of the observed tracks to positrons now follows.

Between August 2, 1932, when the first of the photographs described above was taken, and the completion of the *Physical Review* paper (received by the journal on February 28, 1933), Anderson obtained 15 clear cases of photographs showing

positive particles penetrating the lead, none of which can be ascribed to particles with a mass as large as that of a proton, thus establishing the existence of positive particles of unit charge and of mass small compared to that of a proton.[19]

While the data was sufficient to show that a new particle had been observed, it was not adequate to determine its exact mass and charge. The data did, however, provide striking upper limits of 'less than twice that of the negative electron' for the charge, and 'about twenty times that of the electron'[20] for the mass. It should be recalled that the accepted value for the mass of the proton was 1,850 times that of the electron.

There are several features of this case that will tend to make philosophers uncomfortable with the claim that Anderson was observing positrons, but one that has not yet been discussed stands out: the explicit role that inference played in Anderson's determination of what he had observed. I want to suggest, however, that there is no incompatibility between the claim that Anderson observed positrons, and the claim that he inferred their existence and their properties from the traces on his photographs. For the function of inference here is to bring the available background knowledge to bear on the case at hand. As was noted above, all epistemically relevant observation requires the application of background knowledge, and the fact that, in familiar cases, this often occurs without any explicit inference, permits us to forget the role that background knowledge plays even in our everyday observations. For example, it takes no conscious reasoning to observe my son when I see a dim figure moving towards me through the twilight, or to observe the state of my neighbor's muffler on hearing a particular raucous noise; yet neither of these observations would be possible without deploying a great deal of knowledge. My point here is not to suggest that these cases involve some kind of unconscious or implicit inference, but only to insist that, however it enters in, these observations are dependent on my background knowledge, and that the more reliable this background knowledge, the greater the reliability of the observation. Explicit inference is simply one way in which background knowledge enters into observation.

*3. Neutrinos*

I want to consider one more example, what physicists consider to be the first observation of free neutrinos. Unlike Anderson's unexpected discovery of positrons, neutrinos had been postulated for theoretical reasons, and their properties were well understood before the observation in question. Thus this observation was the result of a search guided by the known properties of the particle being sought, and the observational procedure was considerably more complex than that involved in Anderson's case.[21] I will summarize the main features of the procedure.

Since theory predicts that neutrinos are emitted during nuclear decay, the detection equipment was set up near a nuclear reactor which was expected to produce the enormous number of neutrinos and antineutrinos required to insure a few interactions with the detector. The targets at the heart of the detector were a pair of tanks of water in which cadmium chloride ($CdCl_2$) had been dissolved. These targets were sandwiched between tanks containing a liquid scintillator and a large number of photomultiplier tubes. Occasionally an antineutrino from the reactor will interact with a proton in the water detector yielding a positron and a neutron (inverse beta decay). The positron will encounter an electron in the water target and the two antiparticles will mutually annihilate, yielding a pair of characteristic gamma pulses. Almost simultaneously, but slightly later, the neutron will be captured by a cadmium nucleus, and this interaction will also yield a set of characteristic gamma pulses. The gamma pulses generate flashes of light as they pass through the scintillator, the light is detected by the photomultiplier tubes, and the output from these tubes is fed into the experiment's electronic circuitry for evaluation. The electronics were designed to detect the height of the characteristic electron-positron annihilation pulses and the neutron capture pulses, as well as the time lag between these pulses. In addition, the pulses were displayed on a oscilloscope and the oscilloscope images were photographed for examination later; this provided an extra check for eliminating spurious events. Any case in which an electron-positron annihilation pulse was followed by a neutron capture pulse within nine microseconds was considered to have resulted from the interaction of an antineutrino and a proton, and was thus considered to be an observation of an antineutrino.

Perhaps the most striking feature of this example is its technical and theoretical complexity. With regard to the latter, I want to emphasize one point. Since anti-neutrinos were actually detected, the experiment could not have been designed before physicists recognized the existence of antimatter, and understood its relation to ordinary matter. Anderson's discovery of the positron stands as the first observations of antimatter, and these observations, along with Dirac's almost simultaneous theory of antimatter, provided a precondition for the neutrino observation. In other words, this particular neutrino observation could not have been made until scientists were confident that antimatter existed and could be observed. This, of course, is part of the normal process by which previous scientific results provide a basis for new results. The future history of neutrino observations is striking in this respect, since neutrinos rapidly changed their status from that of particles that had been postulated but never observed, to particles that are routinely detected in the course of observing other items.[22]

With regard to technical complexity too, I want to emphasize one point. Complicated as the observation may be, it was based on a body of physical knowledge that had been developed as physics developed, and it was as reliable as that knowledge. Again we have an example of the way in which new science is built on the results of earlier science, a process that, in the case of observational procedures, has been going on continuously since the first use of a simple magnifying glass.

## III. Conclusion

No doubt many philosophers will still be uncomfortable with the claim that we observe such things as positrons and neutrinos, and will suggest that scientists are using language loosely when they speak of observation in such contexts. Now little of importance hangs on our decision to apply or withhold the label 'observation' in these cases – as long as we properly understand the role that these procedures play in science, and their relation to what has historically been referred to as 'observation'. The fundamental epistemic role of observation is to gain reliable information about the items we are studying, and in so far as we are dealing with physical objects that exist apart from us, observation *always* requires that we extract information about the item that interests us from some effect of a causal processes that involves that item. As our knowledge and instrumentation have become more sophisticated, scientists have developed means of collecting information about items whose existence was not suspected a short time ago, and better ways of collecting information about items we have long known about. The result is that scientists engaged in complex observational procedures are pursuing the same epistemic goals as earlier scientists who used much simpler procedures, and these newer procedures are a consequence of the development of science itself. No doubt earlier scientists did not envisage cloud chambers, or electron microscopes, or CAT scans when they spoke of observation, and there is some point to the suggestion that a contemporary scientist who talks of observing a virus with an electron microscope is not making use of quite the same concept of observation as a seventeenth-century scientist. But this is as it should be. From a naturalistic point of view, concepts are human creations that are developed in particular situations. Contemporary scientists have discovered means of gathering information about the world that were not imagined by earlier scientists, and we should not expect that concepts which were tailored to those earlier procedures should be wholly adequate for discussing newer procedures. But we are not dealing with an utterly new concept of observation either. Rather, the relevant concept of observation for contemporary science has developed out of the older ones, and Shapere's characterization of the contemporary concept of observation as a 'rational descendant'[23] of the older one is particularly apt.

## Notes

1. I want to thank Dr. Donna Baird for comments on an earlier version of this paper.
2. The term derives from Quine (1969), but other contributions include Churchland (1979), Hooker (1974, 1975), and Shapere (1977, 1982, 1984b).
3. Throughout this paper I will use 'item' as a neutral term to cover entities, properties, processes, or anything else that may be subject to observational study in physical science.
4. In this paper I use the term 'reliable' in the common sense in which it is rougly synonymous with 'trustworthy' or 'accurate', and not in the technical sense that contrasts reliability with validity.
5. See Brown (1985) for detailed discussion.
6. Galileo discusses this phenomenon whenever he discusses his telescopic observations, and applies it to the solution of several objections that had been raised by his contemporaries. His most detailed discussion occurs in *The Assayer,* see Galileo (1960), pp. 314–327. The contemporary account of this phenomenon follows Galileo's model, although the detailed analysis of the way the light fringe is generated is much more complex than anything Galileo could have imagined. See Brown (1985) for details.
7. See Churchland (1979) for a detailed development of this point.
8. Boring (1950), p. 135. I want to thank my colleague Wayne Hershberger for calling this example to my attention.
9. Boring (1950), p. 137.
10. Evans (1968), p. 27.
11. Cf. Anderson (1932) and (1933).
12. Anderson (1933), p. 491.
13. Anderson (1932), p. 238.
14. Anderson (1933), p. 491.
15. Anderson (1933), p. 491.
16. Anderson (1933), p. 491.
17. Anderson (1932), p. 238.
18. Anderson (1932), p. 238.
19. Anderson (1933), p. 493.
20. Anderson (1933), p. 493.

21. Cf. Reines and Cowan (1953), Cowan, *et al.* (1956), Reines (1979).
22. Cf. Shapere (1982) for a detailed discussion of one example.
23. Shapere (1982), pp. 507–508, (1984a), p. xxvii.

## Bibliography

Anderson, C.D. (1932), 'The Apparent Existence of Easily Deflectible Positives', *Science* 76: 238–239.
Anderson, C.D. (1933), 'The Positive Electron', *Physical Review* 43: 491–494.
Boring, E.G. (1950), *A History of Experimental Psychology.* second edition (New York: Appleton-Century-Crofts).
Brown, H.I. (1985), 'Galileo on the Telescope and the Eye', *Journal of the History of Ideas*, 46: 487–501.
Churchland, P. (1979), *Scientific Realism and the Plasticity of Mind* (Cambridge: Cambridge University Press).
Cowan, C.L., *et al.* (1956), 'Detection of the Free Neutrino: A Confirmation', *Science* 124: 103–104.
Evans, D.E. (1968), *Observation in Modern Astronomy* (New York: American Elsevier Publishing Company).
Galileo, (1960), 'The Assayer,' in *The Controversy on the Comets of 1618,* S. Drake and C. D. O'Malley, trans. (Philadelphia: University of Pennsylvania Press).
Hooker, C.A. (1974), 'Systematic Realism', *Synthese* 26: 411–497.
Hooker, C.A. (1975), 'Philosophy and Meta-Philosophy of Science: Empiricism, Popperianism and Realism', *Synthese* 32: 177–231.
Quine, W.V. (1969), 'Epistemology Naturalized', in *Ontological Relativity* (New York: Columbia University Press).
Reines, F. and Cowan, C.L. (1953), 'Detection of the Free Neutrino', *Physical Review* 92: 830–831.
Reines, F. (1979), 'The Early Days of Experimental Neutrino Physics', *Science* 203: 11–16.
Shapere, D. (1977), 'What Can the Theory of Knowledge Learn from the History of Knowledge?' *The Monist* 60: 408–588.
Shapere, D. (1982), 'The Concept of Observation in Science and Philosophy', *Philosophy of Science* 50: 485–525.
Shapere, D. (1984a), 'Introduction', in *Reason and the Search for Knowledge* (Dordrecht: D. Reidel).
Shapere, D. (1984b), 'The Character of Scientific Change' in *Reason and the Search for Knowledge* (Dordrecht: D. Reidel).

# Realist Methodology in Contemporary Genetics*

RICHARD M. BURIAN

*Virginia Polytechnic Institute and State University. Blacksburg, Virginia*

## 1. Introduction and stage setting

The present paper is intended to support three main claims. All three, I believe, represent minority views among historians and philosophers of science; they are, therefore, probably of general interest. Boldly stated, the claims to be defended are:

1. Neighboring subdisciplines within a single science need not be methodologically uniform, at least not with respect to realist vs. non-realist methodology.
2. There is an important distinction between realist methodology on the one hand and ontological (semantic) realism and epistemic realism on the other. One piece of evidence for this claim is the fact that, among working scientists, there is no simple correlation between the use or rejection of realist methodology and the acceptance or non-acceptance of ontological or epistemic realism with respect to the leading entities or theories of concern to them in their (sub)discipline.
3. In some very weak sense, methodological realism (which is allied to what Laudan has called 'axiological realism'[1]) is prior to semantic or epistemic realism. The reason is that it is not possible to ground either epistemic or semantic realism adequately without working through the concrete ways in which putative theoretical (or deep sub-structural) entities might enter into concrete experimental and natural situations. Thus one's way of working (methodology) must presuppose the existence of the very entities about whose existence one is uncertain if one is to accumulate and assimilate the evidence that might establish their existence.

Before starting to work on these claims, I shall explain a couple of my prejudices. This will help to clarify the structure of the argument and provide the reader with a firm footing for criticising the position I am defending.

To begin with, what is realism, what are these different sorts of realism? In the first instance, realism is a thesis or attitude toward (putative) *entities, events, processes,* or *states of affairs,*[2] and only secondarily a thesis or attitude about theories. Ontologically, a realist attitude with respect to some supposed entities holds that they ('really') exist whether or not our characterizations or descriptions of them are exactly right. On this sort of account, a scientific realist attitude about electrons or

*Nancy J. Nersessian (editor), The Process of Science. ISBN 90-247-3425-8.*

genes or what-nots would *not* be committed to the literal truth of our theories of electrons or genes or what-nots. Rather, it would be committed to claims along the following lines: a significant portion of our theories dealing with genes, electron, or what-nots, *as standardly applied,* deal with cases or situations in which there are causally efficacious distinctive units in nature. These units, however they should properly be described in the long run, are picked out by use of the relevant labels and are adequately isolated and singled out by the available theories and techniques. That is, these entities belong to 'the furniture of the world'. In philosophy of science, what I call variously *ontological* or *semantic* (scientific) *realism* (about a group of entities, particularly of so-called theoretical entities) is just the thesis that those entities really exist, that the relevant theoretical terms refer to genuine entities, that those entities belong to the spatio-temporal or the causal order.

What I call *methodological realism* holds, roughly, that workers in the relevant discipline(s) should construct their experiments, calculations, and tests by use of the presupposition that there are entities corresponding more-or-less to those which, *prima facie,* are identified in the theoretical portions of the relevant theories. There is also a further commitment to the idea that one of the primary follow-up jobs of the scientist is to get a better 'fix' on those entities. This can be done in various ways – for example, by improving our means of localizing them, by increasing our understanding of their causal and historical roles in interactions of interest, and by increasing the accuracy and suitability of the conceptual apparatus by means of which we describe them, their salient features, or the interactions into which they enter. Methodological realism resembles Laudan's axiological realism by making the treatment of so-called theoretical entities *as if they were real* into a central and salient value of working scientists.[3]

What I call *epistemic realism* is the position that our accounts of the entities in question are sufficiently matched to their 'true natures' to allow us to justify the claim that the entities in question ('really') exist *as described.* One aim of inductivist, hypothetico-deductivist, and similar theories of scientific inference is to provide the apparatus to justify such claims. The crisis of confidence affecting many scientific realists stems, in part, from the problematic status of all such pretensions to have *the* truth (or enough of it) well enough in hand to justify inferences to the specific character of the theoretical entities with which we deal. Epistemic realism, thus characterized, is a stronger thesis than ontological realism; it holds not only that there are entities of the sorts in question, but that our descriptions of those entities are correct and that our specific claims about them are justified in detail.

Much of the literature on scientific realism is skew to the positions that I have been describing,[4] for it treats such realism as a set of theses or attitudes about *theories* and the truth of *theories.* Many of the anti-realist arguments in the literature rightly attack these orthodox formulations of scientific realism. As soon as the viability of scientific realism – or the existence of electrons or genes or what-nots – is thought to turn on the literal truth of the theories which deal with electrons or genes or what-nots, trouble arises. Our theories simply aren't good enough. This trouble is amplified drastically by the admixture of the standard accounts of the meaning of theoretical terms and of the Tarski theory of truth which, together, entail something like the claim that *any* error or falsehood in the fundamental principles, axioms, or postulates of a theory means that all of the descriptions of the 'theoretical entities' in question are literally false. If anything like this holist picture is adopted for the semantics of theories, then it becomes utterly unreasonable to advocate epistemic realism. Such holism means that specific claims about the role of a theoretical entity in a given reaction or interaction must count as false if the general theoretical characterization of that entity is even partly mistaken. The standard involved is thus an all-or-nothing standard; perfection or falsehood. Unless one has – and can prove that one has – the perfect theories required by this sort of semantic realism, it is impossible to justify any claim to have even referred to theoretical entities; all putative references to those entities are embedded in false statements.

These last few remarks should motivate the treatment of realism as a thesis about entities rather than theories.

A second prejudice critically affects the proper use of case studies. What I find worrisome is the *improper* globality of the typical arguments about (and versions of) scientific realism. Part of the burden of this paper is to argue that there are no simple correlations between methodological, semantic, and epistemic realism. More interesting, perhaps, is a background assumption which I shall employ, but which will not be at the center of my argument. One cannot, I believe, perform a strictly formal semantic analysis on (some formulation of) a theory and *on that basis alone* determine the set of entities to which realists who work with theory would or should be committed. To use a straightforward example from Newtonian mechanics, it is at least arguable that masses, mass-points, and spatiotemporal points do not all have the same status. (I, for one, think that Newtonian mechanics need only be committed to the reality of masses, but that there is a much stronger case for commitment to the existence of spatiotemporal points than there is for the existence of mass-points, which are flagrant idealizations.)

For an argument to this effect to get off the ground, much more than formal-semantic analysis of Newtonian mechanics is required. After all, in appropriately chosen formulations of Newtonian mechanics the terms that pick out all three sorts of entities look, grammatically speaking, like designating terms. Such differences in status as there are need not be, and typically are not, reflected in grammar. Yet traditional arguments over realist vs. anti-realist interpretations of such theories as Newtonian mechanics suppose that some such tool as formal semantics will be sufficient – and sufficiently general – to work out the correct resolution of all such disputes. This expectation is wholly unreasonable. To settle the issues about masses, mass-points, and spatiotemporal points, we must look into the procedures for isolating, identifying, and individuating entities of each of these kinds, and ask whether they are adequate to pick out real things or whether they automatically bring with them a corresponding denial of reality (as the average family, with its 2.3 children per family does). The great complexity and variety of procedures for isolating, identifying, and individuating entities and properties employed in science make the application of global tools of analysis at best very difficult and at worst utterly implausible. Local considerations play a very critical role in all serious cases in which the viability of a realist interpretation is at stake. These must be worked out on a case-by-case basis in light of the best available knowledge as to how to get at the various (putative) entities.

Return to the Newtonian example. Notice the lack of any procedure for identifying mass-points independent of the taking of certain limits as one shrinks the boundaries of masses (while, *per impossibile* retaining the entire mass). Add the theoretical demonstration that the outcome of certain calculations is unaffected by substitution of the mass-point values thus derived for the calculationally intractable values for distributed masses. The result is that point masses are theoretically, though perhaps not practically, dispensible idealizations. No argument this simple works for spatiotemporal points. And when it comes to masses, the argument that commitment to those is wholly indispensible turns, on my account, on a detailed examination of the procedures by means of which masses are picked out and by means of which the relevant property (or properties) is delimited from others. On this approach, it follows from the details of the relevant history of work in physics that there is no Kuhnian problem of incommensurability regarding mass: the property in virtue of which well identified bodies of various scales exhibit inertial resistance *is* picked out by a vast number of uses of the Newtonian term 'mass', but that property does not behave in exact accordance with Newtonian formulae.

Such a story, if it goes anywhere at all, does so in virtue of what I have been calling 'local' considerations regarding the detailed ways in which one mass is distinguished from another and in which the property of mass is distinguished from other properties of material bodies. In particular, the story I have been telling *cannot* be built by using the traditional apparatus of logical analysis of theoretical claims, and especially not by an analysis which claims that a Newtonian mass must, at the very least, be something which satisfies the axioms of Newtonian mechanics.

Though this point requires case-by-case testing, I think that it generalizes pretty well. In the various sciences different experimental, observational, and theoretical procedures are employed as the means of identifying and individuating particular 'theoretical' entities and properties. Any reasonable assessment of the merits or demerits of a realist's interpretation of those entities and properties must take fairly full account of the merits and demerits of those 'local' procedures in the full light of background knowledge, broadly conceived.

For such reasons, one should not expect much uniformity within or between sciences. Case studies are critical to any serious attempt to grapple with the nest of problems associated with realism. Frustratingly, the answers which case studies provide will have a relatively limited application, allowing some sort of estimate of the reasonableness of a realist stance with respect to a certain range of cases and entities, but no general answer as to whether science (in general) should be given a realist interpretation. This is frustrating, but sensible. When one thinks back on it from the perspective put forward in the present paper, the whole idea that a relatively abstract and formal analysis of the logic or structure of scientific theories plus an abstract characterization of scientific methodology should be sufficient to provide the footing for a general interpretation of theories seems bizarre. After all, it is in the details of its dealings with real objects that science provides knowledge of the world (to the extent that it does so), and these dealings are simply not captured by the structure of finished theories plus the general principles (if any prove viable) of scientific methodology.

Before turning to the cases to be discussed, one last preparatory note is needed concerning their importance and value. I will be discussing rather recent work in a couple of subdisciplines of genetics. Such work is of interest to philosophers for many reasons. For one thing, as I hope to show, genetics is a field in which realist methodologies predominate while realist and non-realist ontologies are employed distinctively in different subdisciplines. For another, it is a discipline which has been undergoing quite dramatic change in ways which seriously undermine traditional semantic and epistemic realism regarding older genetic theories, and, to a lesser extent, regarding some of the entities to which those theories purportedly referred. For a third, with respect to the present issues by far the best way of interpreting the recent turmoil in molecular genetics is to view it as revealing what happens when realists of one persuasion (molecular geneticists) muster sufficient evidence to show that realists of another persuasion (traditional Mendelian and transmission geneticists) must seriously alter, or perhaps even abandon, their account of the entities whose existence they thought they had established. A realist stance with respect to the older entities (traditional genes) is in the process of being replaced by a realist stance regarding a different class of entities (the extremely diverse entities of molecular genetics). This process is of considerable interest to those of us concerned with conceptual change in science. Yet, fourth, in such neighboring (sub)disciplines as population genetics and quantitative genetics, this sort of replacement is not occurring. The *methodological realism* which is intimately allied to the practice of modelling in both population and quantitative genetics does not carry with it a strong commitment to ontological or epistemic realism regarding the fundamental units employed in the various models. Accordingly, the main effect of the upheaval occurring in molecular genetics on population and quantitative genetics is to broaden somewhat the class of models which the latter disciplines take into account, but not to change the foundations or the ontologies of these subdisciplines.

If all of this is correct, case studies dealing with these matters provide an ideal vehicle for probing the strengths and limitations of the sorts of localized realism which I have been suggesting. Let us employ them, then, to examine the complex interrelations between the methodological, epistemological, and ontological moments in a realist stance.

## II. On some differences among molecular, population, and quantitative genetics

Population genetics studies the changing composition and the genetic structures of populations (and, in certain respects, of the organisms out of which those populations are constituted) over time. Quantitative genetics studies the inheritance of traits influenced by many genes or genetic factors, especially those like height and milk production that are readily quantified. Both depend very heavily on the use of mathematical modelling, and both typically get at the genetic structures in which they are interested by examination of the *phenotypes* of the organisms in question. Thus, for example, the classic case of heterozygote superiority, sickle cell anemia in humans, was first worked out by population techniques without any specific knowledge of the particular genetic defect which caused red blood cells to sickle.[5] What was noticed was that the sickling trait was inherited in accordance with a classic Mendelian pattern and that it was present in certain Mediterranean and African populations at far higher rates than would be expected given that its sole (known) effect was to cause sickle cell anemia (effectively a lethal condition) when it was present in homozygous dosage.

When it was noticed that the disease was present primarily in populations which resided where falciparium malaria is (or was recently) present, two steps were taken. The first was to ask whether, *de facto,* individuals who had the sickling gene in single dose were better able to resist malaria than individuals who did not carry the gene. (The answer to this question proved positive.) The second was to develop for application to this case the mathematical models of the spread of a gene which, though lethal when homozygous, conferred an advantage when heterozygous. Among the things that one could do with such models was the following: use the prevalence of sickle cell anemia where malaria is severely endemic to establish a value for the selection coefficient favoring the heterozygote over the 'wild type' and the sickle cell homozygotes. Given the resultant assignment of selective values, one could then predict the change in the selective advantage for the heterozygote, and hence the rate of loss of the sickle cell gene, when the population encountered conditions in which malaria was less frequent or nonexistent. Although it is quite difficult to find a clean case in which it can be established that, with respect to the relevant parameters, a population started in a given initial condition, has been transported to (or otherwise encountered) a well-defined alternative condition and remained sufficiently intact for the modelling assumptions to apply, there have been a few instances in which this has occurred (e.g., with respect to certain populations of American blacks who were transported from malarial to wholly non-malarial environments under adequately known population-structural conditions); in all of the instances with which I am familiar, the model predictions have held up extremely well.

Furthermore, in this instance, additional, independent corroboration of the principal results of the model has been obtained via the subsequent detailed elucidation of the molecular mechanisms of sickling. The mutation causing sickle cell is a single point mutation in the gene encoding hemoglobin. The hemoglobin of an organism homozygous for this mutation crystallizes under low oxygen tension, thus distorting the red blood cells into the characteristic and useless sickled configuration. In heterozygotes the oxygen content of red blood cells is too low to support the falciparium parasite but not low enough to cause important physiological disadvantage to the carrier of the trait. These results integrate the population genetic claims into a larger framework and provide a kind of consilience with those parts of the molecular picture which deal with related matters.

Although this is an unusually straightforward case of population genetic modelling (and also, evidentially considered, an unusually clean case), it provides enough detail to allow me to make some preliminary points about the methodology of modelling in population genetics. The central point can be put in a nutshell: population genetic models employ a kind of bookkeeping that keeps track of the changing distribution of genetic units under particular assumptions about the initial distribution of genetic units, the rules governing the transmission of those units from generation to generation, and their selective values under particular environmental circumstances. There is also a secondary bookkeeping correlating the distribution of observable states (phenotypes) with the distribution of genetic states. The genetic state of the initial population, insofar as it is defined or understood, is inferred from the initial distribution of phenotypes in the population.

*This procedure makes no sense unless, while using the model, the genetic traits are treated as if they caused the phenotypic states.* (If they were mere correlates of those states, the laws governing their transmission would not differ from those governing the transmission of phenotypes.) But the procedure is not wedded in any full-blooded way to the claim that the *genetic* states are properly or correctly described within the model. Anyone who is unhappy with this claim should read Lewis Stadler's posthumous valedictory article, 'The Gene'.[6] This article, though not directly concerned with population genetics, explores the limitations of Mullerian-Morganian genetics in which 'a change in a gene can be recognized only by its effects' (p. 813). Stadler's key point is that in classical transmission genetics genes are not *intrinsically* characterized (e.g., in terms of their molecular structure) and are, at best, imperfectly localized as falling within particular regions of particular chromosomes. Accordingly, he provides a detailed foundation for the claim that neither transmission nor population genetic techniques can adequately discriminate among multiple alleles, gene clusters, changes of gene position, interaction effects, changes in gene expression due to extragenic factors, and so on.

On my account, what matters in population genetics is simply the rules describing the transmission of the genes, the values of the parameters affecting gene transmission, and the rules of correlation between distributions of genes and distributions of phenotypic states. In order to save the appearances, population genetic models can (and frequently must) be made quite complicated – for example by adding peculiar

rules of gene transmission in order to handle various mating systems or the phenomenon of meiotic drive (a bias in favor of one allele over another in the formation
of sperm or eggs). But this does not affect the fundamental point that the treatment
of the presumed genetic units in such models (Stadler calls them 'hypothetical
genes') is properly understood as algebraic shorthand for *whatever* behaves in
accordance with such-and-such transmission rules and has distributions which are
correlated in thus-and-such a way with distributions of phenotypic traits. Since
application of the algebra only makes sense if the distribution of phenotypic traits is a
consequence of the underlying distribution of genetic traits, the use of such models
requires a sort of methodological realism, but does *not* require semantic or epistemic
realism.

These points are in good agreement with some arguments that have been put
forward in support of one side in a current foundational debate regarding the status
of fitness ascriptions at the genic level. In spite of the fact that the models of
population genetics typically presuppose that each allele has a definite fitness, the
ascription of particular fitnesses to particular alleles may make no more biological
sense, in general, than the ascription of 2.3 children to particular families makes in
sociology.[7] There is absolutely no need to suppose that the ultimate genetic units,
whatever they may be correspond exactly to the elementary genetic units employed
in population genetic models or that properties such as fitness should be ascribed to
genes in any more serious sense than *average number of offspring* should be ascribed
to particular families. Indeed, as arguments like Stadler's show, the grounds sup-
plied within population and Morgan-Muller style transmission genetics are typically
too weak to justify any such claims about 'hypothetical genes'. In a way this is
fortunate, for it makes sense of the otherwise nonsensical practice of employing
incompatible models for the genetic structure underlying certain phenotypic traits in
different contexts.

The workings of this practice will be somewhat clearer if we consider, briefly, the
structure of quantitative genetics. Quantitative genetics has its origins in plant and
animal breeding, and even now breeding work provides it with many of its problems
and the touchstones for judging how well they have been solved. It is not surprising,
therefore, that the equations of quantitative genetics, unlike those of population
genetics, are predominantly *phenotypic* equations. That is, they are couched in
terms of the increment of change in phenotypic traits like height, weight, percent of
oil in the seed, and so on. In a fairly typical problem in quantitative genetics, the
equations are used to provide an estimate of the number of 'genes' involved – i.e., of
genetic factors that contribute to the inheritance of the trait. These factors are
divided into two groups – those that contribute additively to the trait of concern and
those whose contribution is non-additive. (The latter group is often subdivided
further.) The allocation between additive and non-additive heritable variation is
critical in breeding contexts, for it determines the choice of a system of mating. The
allocation is adjudged satisfactory if it yields correct predictions regarding the

optimal mating scheme for altering or preserving the desired traits. Estimates of the number of genes and the proportion whose effect is additive are largely (re)descriptions of the pattern of inheritance involved; they do not seriously pretend to describe the underlying mechanism of transmission or the units of heredity whose presence, absence, or variation is responsible, in context, for the phenotypic differences of interest to the breeder or the geneticist.[8] To put it briefly, the equations in question would be described by physicists as 'phenomenological'. Accordingly, when the concern is, in this sense, phenomenological, the models of quantitative genetics will be counted as satisfactory even when they disagree, interpreted literally, about the mechanisms underlying a particular case. What the models *must* do is to preserve the phenomena, i.e., they must correctly describe the transmission of the traits of concern.

This account shows that it is often the case that the use of incompatible models, envisaging quite different numbers of genes underlying a given phenotype, will cause no difficulty in quantitative genetics. In particular, there would often be no perception of conflict or of a need to determine which is 'the' correct number of genes bearing on the phenotype; what is at stake is the pattern of quantitative variation and its transmission in various cases and not the details of the mechanisms that bring that pattern about. Such 'phenomenological' concerns are, of course, not unique to quantitative genetics though they are, I believe, more common there than in population genetics. To this extent, conflicting models are less problematic in quantitative than in population genetics. To this extent, the former is less committed to methodological realism (the presupposition that the real world contains entities matching more-or-less one-to-one with the genes of the relevant models) than is the latter.

The situation of molecular genetics is strikingly different. While the *epistemic* status of claims about the various units of molecular genetics is at least arguable, molecular genetics is much more deeply committed to *semantic* (and hence also methodological) realism than population and quantitative genetics are. This is so because molecular genetics offers (or at least attempts to offer) *intrinsic* characterizations of the fundamental genetic units with which it deals. It does not base its characterization of genetic entities and their properties on second order phenomena, such as the consequences, in context, of the presence or absence of those entities, but attempts to describe the first-order properties of exactly localized entities (e.g., strings of DNA) and to provide the detailed mechanics of the interactions into which those entities enter. Rather than characterizing the relevant entities by virtue of their functions or of the effects that they (or their distributions in a population) bring about, it attempts to derive the functions and effects from an analysis of their structure and of the context in which they occur.

The structure-function paradigms involved here are ultimately mechanical in a rather strong sense; their pattern (in spite of the fact that they are far more complex than their paradigm) is parallel to the one in which the structure of a ball-bearing and its socket explains the bearing's function of mediating nearly friction-free sliding contact between surfaces which would otherwise bind. Thus, the ability of the hemoglobin molecule to trap oxygen molecules in its four heme pockets should prove derivable from the conformal arrangement of the molecule plus a very full knowledge of the conformation of the various molecules it contacts under physiological conditions. Similarly, the functions (and side effects) of the palindromes in various regulatory and transposable DNAs and RNAs should prove to be derivable from the fact that those palindromes inevitably create a characteristic stem and loop shape under physiological conditions and the knowledge that that shape creates a closed loop of DNA or RNA which, under rigidly specifiable circumstances, can be processed in characteristic ways. (Among the relevant processing is that which excises the loop while rejoining the remainder of the molecule and that which transports the loop so as to join it up preferentially with particular sequences of nucleotides wherever they happen to be exposed on the cell's complement of DNA.) There are, of course, innumerable complications, many of them of considerable interest. For example, many interesting cases involve rate competition in such a way that the relevant mechanical descriptions, in context, yield transition probabilities rather than deterministically controlled reactions. But the complications can be accounted for quite well within the confines of the account sketched here.

The central point is not that it is impossible to provide an instrumentalistic or a van Fraassenian interpretation of molecular genetics, and it is certainly not that formal constraints of some sort enforce a semantically realist interpretation of molecular genetics. It is rather that the structure-function models being used here are strikingly similar to those used in animal mechanics in determining the functional consequences of the differences between, say, a horse's and a rhinoceros's leg. A fairly full understanding of those models *as used in real circumstances* makes a non-realist interpretation of them, and of the allied 'theoretical' structures of DNA that they employ, utterly implausible. A central part of my case turns on the fact that the means of localizing and isolating the molecules and of identifying the molecular transitions and configurations at stake are largely independent of the particular theoretical issues which remain under debate. This allows scientists to identify the entities whose structure and function is to be analyzed independently of the details of the analysis in much the way that one can identify a horse's or a rhinoceros's leg independently of a particular analysis of the leg's structure.

The identification and localization of the relevant units is, to be sure, much more indirect and technology-dependent in molecular genetics than in animal mechanics – in the former case it may involve electron microscopy, density centrifugation, autoradiography, and sophisticated electrophoretic analysis. It is much easier to identify and examine legs or lungs than it is to identify and examine transposons or specific subsegments of the fourth chromosome's or the mitochondrion's DNA. But this does not mean that the isolation of the molecular entity or change whose structure(s) and function(s) are to be analyzed is viciously theory-dependent.

### III. Some comments on a realist interpretation of molecular genetics

Because of length restrictions, my treatment of the relationship between current molecular genetics and older theories of the gene must be brief. In a paper currently in progress, I explore the consequences of the fact that the new molecular genetics deals with a bizarre variety of genetic units – split genes, insertion sequences, transposons, gene families, gene scrambling mechanisms, systems for generating controlled somatic rearrangement of genes, and so on – which simply do not correspond with the sorts of entities that traditional theories recognized as genes. Equally startling is the fact that, when genes are defined functionally, the topology of the genome is not strictly linear. The DNA coding for a particular protein, though linearly ordered, typically is not contiguous and the interspersed segments occasionally perform independent functions. The DNA coding for a particular protein produced by the adult may have been produced by systematic scrambling of the DNA present in the zygote. (Incidentally, in such cases the information coding for that protein was not present in the zygote, showing that the genetic information of organisms is not wholly determined at birth.) Again, in extremely rare cases, the information coding for a single protein may be encoded partly in nuclear DNA, partly in mitochondrial DNA. The list of non-linearities goes on; the topology of the genome simply does not correspond to the neat linear topology which Morgan and Watson and Crick would have led us to expect.

Indeed, at most about 10% of the DNA enters into classical genes (or, rather, their contemporary descendents). If one defines genes *functionally,* these results do not undermine the claims that genes exist. In particular, since, in practice, that is precisely how genes were defined (recall the quotation from Stadler about genes being defined by their effects), we can (by and large, with only specific exceptions) be assured of the referential continuity of such phrases as 'the gene for red eyes' or 'the sooty locus'. There *are* genetic units which were picked out by these labels and the same units are still referred to by the use of those labels. But if one accepts a traditional philosophical account according to which a gene is an entity which, whatever else it does, behaves in accordance with the fundamental postulates of some theory of the gene, then one is forced into the position of denying that there are any – or at least many – such things as classical genes. In short, genes (at least by and large) do not behave in accord with the traditional theories, and they include a bizarre diversity of molecular entities which do not accord even approximately with the criteria of individuation and identity which those theories would have placed on genes. If one is required to use the fundamental tenets of the older theories as the basis for identifying genes, then one is forced to say that molecular biology has shown that there are virtually no genes and that the few that there are play a rather modest role in the genetics and evolution of the vast majority of organisms.

If, in contrast, one considers genes to be those entities which *de facto* were isolated or referred to by the techniques and practices of traditional geneticists, then the proper description of the situation is that genes turn out to be vastly different than we thought they were, both intrinsically and, to a lesser extent, in their topology and modes of interaction. It is not critical which terminology we decide to adopt, but it is a fact of fundamental importance that there is massive referential continuity in the face of enormous theoretical disagreement and error. This fact deeply affects the proper understanding of the biology involved and the proper execution of meta-scientific analyses of the theoretical changes in question. The claim of referential continuity with respect to such (previously mischaracterized) theoretical entities as genes, a claim which I have been suggesting is the key to the proper understanding of contemporary molecular genetics, is, of course, one which belongs to the very heart of a semantic-realist account of theoretical entities and, perhaps, of theories.

It need not, of course, have turned out that there was such continuity. It is only if we have done reasonably well in picking out the causally relevant factors in certain interactions that we can then successfully reconceive how those factors are con-structed and how they interact with one another and with the relevant features of their environment. There is, of course, plenty of room for argument over the degree to which the transition from transmission to molecular genetics fits the mold just described. I am prepared, however, to argue that it fits pretty well and that the mixture of referential continuity and conceptual incongruity which are involved here make this case a useful vehicle for exploring the sorts of progress that result from major revisions in our knowledge of the 'theoretical' entities underlying a well-studied range of phenomena. It is to be hoped that the present case is typical enough to illuminate one major avenue of theoretical progress in the development of so-called deep-structure theories. If so, the general approach employed here will provide useful apparatus for dealing with many cases of interest.

### IV. Some conclusions

This paper tells only part of the story. Nonetheless, the general lines of the position being developed here should be fairly clear. There is no reason to expect closely related scientific (sub)disciplines to be uniform in the methodologies they employ or in the degree of epistemic or semantic realism to which they are committed. Individual sceintists will differ widely in these regards, but the biases and tendencies in neighboring subdisciplines will often point in different directions. Some of these tendencies are of considerable importance. To the extent that a theoretical discipline offers an *intrinsic* characterization of the theoretical entities with which it deals, to the extent that it offers extratheoretical means of locating or individuating those entities (again, intrinsically, rather than by their effects), to that extent there will – or should – be a tendency for its practitioners to subscribe to semantic realism. The degree to which epistemic realism with respect to the entities identified by the relevant theories is justified does not turn solely on the evidential status of the fundamental theory (or theories) involved, but also on the extratheoretical procedures for fixing the reference of the relevant theoretical terms. The evidential status of such entities should be evaluated both relatively and absolutely: How well does the theory in question do in comparison with its competitors? How reliable are the practices for picking out and/or localizing the theoretical entities in question? Is the best of the available theories in the discipline robust and reliable or fragile and poorly supported? To what extent can the means of picking out the theoretical entities in question be made independent of the details and idiosyncracies of the leading theory or theories dealing with those entities? It is no surprise that people who work in different corners of a discipline, people who work with different techniques, people who are exploring different ranges of questions on which fundamental theory is brought to bear, will have different answers to such questions as these.

Yet all is not chaos. If the analysis developed here is on the right track, one can, for example, supply a pretty strong justification for the conclusion that in spite of the commitment of population genetics to methodological realism, neither semantic nor epistemic realism with respect to genes is justified on grounds of the results of population genetics *taken by themselves*. That is, population genetics by itself is not powerful enough to justify either the claim that there are genes (a claim that *can* be justified on other grounds), nor that genes proper are well described by population genetic theory. When one takes intertheory relations into account, the situation is altered. An interesting case could be made for the appealing view (which, however, I am not fully prepared to defend) that before the recent flourishing of molecular genetics, transmission genetics offered a reasonable basis for semantic realism with respect to genes, but the epistemic difficulties (elaborated by such internal critics as Stadler and Goldschmidt) were sufficiently great that one could not justify any intrinsic characterization of genes.

To put the point using less jargon, transmission genetics and the experimental procedures with which it was allied genuinely picked out genetic units and structures which were causally relevant to the patterns of inheritance that genetics sought to explain; its descriptions of those units, however, were systematically mistaken, and seriously so. And the evidence for the view that transmission genetics had actually located and described those units accurately, even to a first approximation, was pretty shaky. Unambiguously satisfactory means were not then available to show that the patterns of inheritance which were to be explained by the behavior of the genes were actually caused by inheritance and action of units matching geneticists' descriptions. To this extent, those critics (like Stadler) who argued for an operationalist or instrumentalist interpretation of the gene were justified in claiming that it had not yet been adequately shown that entities of the sort *putatively* isolated by genetic experiments, localized by (for example) chromosome mapping procedures, and described by one or another theory of the gene *really* were present and *really* provided the substructure which accounted for the hard-won phenomena which geneticists sought to explain. In all events, to evaluate the status of the entities described in particular population genetic models one has to examine the relationship between the entities those models postulate and the ones described in transmission (or, currently, molecular) genetics as well as the means for picking out and localizing those entities.

The case in favor of this view depends, to a great extent, on matters of detail about the procedures for identifying, individuating, and locating genetic units. Yet, typically, one must determine whether or not to employ a realist methodology with respect to a group of (putative) theoretical entities on the basis of promisory notes and without a full or comprehensive evaluation of those procedures. Indeed, the division of labor in science makes it unlikely that someone working in a field like population genetics is in a good position to offer a synoptic evaluation of (or to have a broad appreciation for) the procedures by means of which genes (and other fundamental genetic units) are, in general, picked out, isolated, localized, and individuated. This result reinforces my claim that there need be no simple correlation between the use of a realist (or non-realist) methodology and the acceptance of rejection of semantic or epistemic realism with respect to the entities at stake. More interestingly, it also suggests that the status of claims about the fundamental theoretical entities employed within a particular theory or discipline may well turn on the outcome of investigations in related disciplines even though there is no question whatsoever of 'reducing' the entities of one discipline to those of the other. This suggestion I take to be of quite general interest.

## Notes

* Work on this paper was supported in part by an NSF Scholar's Award for calendar 1984. This support is gratefully acknowledged. Preliminary versions of the paper were read in April, 1984, at a Conference on Realist Methodologies at the Center for the Study of Science in Society, Virginia Polytechnic Institute and State University and in April, 1985, at the Pacific Division Meetings of the American Philosophical Association. I have been greatly helped by critical discussion on both occasions, by the comments of John Dupre, and for the suggestions of numerous colleagues, especially M. Grene, J. Griesemer, L. Laudan, R. Richardson, P. Siegel, and B. Wallace.
1. Laudan (1984).
2. Marjorie Grene rightly objects that in most of the text I speak as if theoretical terms refer only to entities. This way of speaking is only for ease of exposition; the sorts of 'entities' referred to in successful theories can just as easily be states of affairs, events, processes, and so on as 'things.' No implications about the ontological categories to which theoretical entities in general belong are intended in my exposition. Furthermore, in many cases it may prove critical to purge our account of the reference of theoretical terms of the 'entity-centrism' of expositions like the present one.
3. Methodological realism also resembles certain features of the 'natural ontological attitude' (NOA) discussed by Arthur Fine (1984), particularly in the 'local' and discipline-specific nuances it allows in the interpretation of theoretical terms and the determination as to which of those terms are referring terms.
4. But by no means all. Among the many recent honorable and interesting exceptions are Cartwright (1983), Hacking (1983), and, in closer contact with the immediate topic of this paper, Rosenberg (1978).

5. This case is discussed in virtually every standard textbook in genetics or population genetics. One of the better brief treatments can be found at pp. 616 ff. of Strickberger (1976). An interesting early treatment of the case in terms of genetic polymorphism is in A. C. Allison (1955).

6. Stadler (1954).

7. Positions of this general sort are found, for example, in Sober and Lewontin (1982) and Wimsatt (1980). Note that with appropriate use of means and variances, one can build models that treat the mean number of children in a 'typical family' (e.g., 2.3) and variances around that mean as the principal independent variables. (Similarly for allelic fitness of 1.2 etc., plus variances around that mean number.) The models built with such independent variables may well prove calculationally adequate. When they do, it is very easy to fall into the trap of treating those variables (as the model must) as causally sufficient to explain the interaction in question. The point at stake here is precisely that a substantial argument, involving issues external to the relevant models, is required to justify semantic or epistemic realism with respect to the 'entities' putatively named by the *prima facie* designating terms of the models. That is, it requires independent argument to justify such claims as those (maliciously chosen) to the effect that there really are families with 2.3 children and that we are justified in the claim that our theoretical apparatus describes them correctly. The success of the relevant models does not, it is obvious, suffice to justify such claims. And in the case of the fitnesses assigned in population genetics, one is dealing with *averages* even when assigning fitnesses to individual alleles, so that fitnesses are, in this respect, precisely parallel to average numbers of offspring. (I owe the following way of putting this point to Bruce Wallace.) In standard symbolism, such fitnesses as $W_{AA}$ are averages, determined by taking the average effect of AA over the various genotypes, phenotypes, and environments in which it occurs. What are called mean fitnesses, conventionally symbolized by $\bar{W}$, are thus averages of averages.)

8. This description applies to a major part, but by no means the whole of quantitative genetics. It is worth insisting that it is not a straw man. Cf., e.g., Section 63 ('The Effective Factor') of Mather and Jinks (1982), pp. 367 ff. For example: 'The various studies of [a certain trait] suggest that at least 15 to 20 genes must be involved in its determination . . . The values found for the number of effective factors are in striking contrast to this expectation, for they are generally smaller and often strikingly small. The reason for this discrepancy is, of course, clear . . . [T]he effective factor is not of necessity the ultimate gene, and the estimate of the number of factors is further reduced by the size of their effects' (p. 367). 'Operationally we can recognize a gene only by the difference to which it gives rise. If the gene does not vary and, in varying, produce a detectable phenotypic difference, we cannot know that it is there. The inference of a gene is therefore limited by the means available for detecting the effect of its variation. If the difference produced in the phenotype is too small to be picked up by the means at our disposal we cannot identify the determinant, though we may be able to detect a group of such determinants of small effect when they are acting together' (p. 368).

'The number of factors into which a chromosome can be divided [by recombination tests] cannot, of course, exceed the ratio its total genetic length bears to the recombination frequency of adjacent factors' (p. 370). ' . . . in biometrical genetics, observations can be made only on linked groups of the genes, on in fact the total genic content of factors whose physical basis is to be found in whole segments of the chromosomes. The properties of the groups of effective factors will reflect not merely the behaviour of each individual constituent gene, but also the mechanical and physiological relations of the genes with one another. And since we have no means of predicting these genic relations in detail, we can learn the properties of the factors only by direct observation and experiment, as indeed we have already begun to do' (p. 376).

In short, without intrinsic characterizations of single genes, there will always be models available in population and biometrical genetics which, though equivalent with respect to all available observations, disagree about the number, location, and linkage of the relevant genes. Because of this, the main thing that modellers are concerned to establish is that their models contain 'effective factors' that are sufficient to save the phenomena.

## Bibliography

Allison, A.C. (1955), 'Aspects of polymorphism in man,' *Cold Spring Harbor Symposia in Quantitative Biology* 20: 239–255.

Cartwright, N. (1983), *How the Laws of Physics Lie* (Oxford: Oxford University Press).

Fine, A. (1984), 'The natural ontological attitude', in *Scientific Realism*, J. Leplin, ed. (Berkeley: University of California Press), pp. 83–107.

Hacking, I. (1983), *Representing and Intervening* (Cambridge: Cambridge University Press).

Laudan, L. (1984), *Science and Values: the Aims of Science and their Role in Scientific Debate* (Berkeley: University of California Press).

Mather, K. and Jinks, J. (1982), *Biometrical Genetics,* 3d ed. (London: Chapman and Hall).

Rosenberg, A. (1978), 'The supervenience of biological concepts', *Philosophy of Science* 45: 368–386.

Sober, E. and Lewontin, R.C. (1982), 'Artifact, cause, and genic selection, *Philosophy of Science* 49: 157–180.

Stadler, L.J. (1954), 'The gene', *Science* 120: 811–819.

Strickberger, M. (1976), *Genetics,* 2nd ed. (New York: MacMillan).

Wimsatt, W. (1980), 'Reductionistic research strategies and their biases in the units of selection controversy', in *Scientific Discovery*, Vol. 2, *Case Studies,* T. Nickles, ed., (Dordrecht: Reidel), pp. 213–259.

# Parsimony and the Units of Selection

ELLIOTT SOBER

*University of Wisconsin, Madison*

## 1. Methodological preliminary

Philosophers of science often express suspicion of the remarks scientists make about scientific method. These pronouncements, so it is said, often are after-the-fact rationalizations that fail to reflect the real epistemological principles that scientists obey in practice. How often have we heard that if Einstein had been the Machian he sometimes said he was, he never would have discovered the special theory of relativity?

There is more than a little truth in this maxim. But as with any maxim, it is important not to exaggerate its correctness. Scientists often *do* have real insights into scientific epistemology. One must listen to what they say and then subject it to critical scrutiny. As with any maxim, once tempered, it verges on triviality.

The above remarks apply with equal, if not greater, force to the remarks that philosophers make about philosophical method. I do not exempt myself from these suspicions, which is why I feel slightly uncomfortable with the editor's request that I preface my philosophical discussion with a discussion of the method I use.

Philosophy of science often has vacillated between an *a priori* normative assessment of scientific theories and a complete rejection of the possibility of normative philosophical critique. When in the former mood, philosophers sometimes produce wholesale condemnations of entire theories or disciplines; evolutionary theory, cognitive psychology, psychoanalysis, Marxist theory – all these and a few more have found themselves written off. Characteristically, such philosophical condemnations involve little or no detailed scrutiny of the content of the theories involved. It is a laughable caricature to equate evolutionary biology with the slogan of the survival of the fittest. It also is characteristic of such wholesale condemnations that they involve the application of some very simple formula that defines REAL SCIENCE. Besides showing the unreality of their pictures of the theories under scrutiny, such critiques often reveal the inadequacy of the philosophical standards as well.

Philosophers drawing back from the excesses of *a priorism* often think that it is not for philosophy to dictate to the sciences. Philosophy must try to understand how science in all its variety actually works; normative critique is impossible, if the results

*Nancy J. Nersessian (editor), The Process of Science. ISBN 90-247-3425-8.*
© *1987, Martinus Nijhoff Publishers (Kluwer Academic Publishers), Dordrecht. Printed in the Netherlands.*

of science are more trustworthy than the deliverances of philosophical reflection.

I would urge that both these methodological prescriptions are misguided. Each rests on the idea that there is a crisp distinction between two monolithic entities called SCIENCE and PHILOSOPHY. Normative a priorism thinks of PHILOSO-PHY as prior to SCIENCE, whereas reactive descriptivism imposes just the opposite ordering. However, if the boundary between philosophical and scientific questions is often blurry, it will be misguided to demand to know which is answerable to which.

This assessment is licensed in theory by a philosophical thesis – namely, the rejection urged by Quine[1] of the analytic/synthetic distinction. In practice, it is confirmed by attention to the details of scientific arguments. A scientific discipline often proceeds without any overt discussion of philosophical issues when it is in a period of relative consensus about aims and methods. However, when a foundational controversy arises, the evaluation of philosophical issues can become a pressing practical concern. In such cases, philosophical questions are not alien to the pursuit of science, but grow organically out of science itself.

It is in such scientific controversies that a philosopher may find a foothold for the pursuit of normative questions that avoids the two extreme positions described above. There is no first philosophy, prior to all of science. But this does not mean that all scientific positions are equally good. An *applied* epistemology, attentive to the details of scientific practice, can hope to evaluate lines of argument and codify principles that are properly called philosophical.

A philosophy rooted in the details of scientific controversy is not limited to matters that actually concern practicing scientists. Philosophy has always aspired to a kind of generality that goes beyond the subject matters specific to any one scientific discipline; there is no *a priori* reason for thinking that this is impossible. Besides being an underlaborer for the sciences, philosophy also encompasses a set of problems that have had and will continue to have an autonomous philosophical interest. Although philosophy may begin with matters of scientific concern, it need not end there.

## 2. Williams' parsimony argument

Parsimony is one of those ideas that surfaces every so often in biology, as well as in other sciences. Some would see it as a pervasive, though usually implicit, methodological rule that constrains all scientific activity. Although this may be true, my present interest in parsimony is at a much less lofty level of generality. I am interested in examining the details of how parsimony has been used in a research area in evolutionary theory. This more detailed scrutiny is useful, I think, in the way it helps expose pitfalls of parsimony arguments and also in the way it leads one to a clearer picture of what is necessary for a justification or refutation of parsimony within a circumscribed theoretical context. Perhaps some general theory of parsimony will someday be described that subsumes such local issues as special cases. But as a working hypothesis, I suggest that the truth about parsimony is all in the details.

In his important book *Adaptation and Natural Selection,* George C. Williams uses parsimony as one of his main arguments against hypotheses of group selection.[2] He discusses numerous cases in which a character has been observed to be common in a population. After describing a group selection explanation that some other biologist had suggested, Williams formulates an alternative hypothesis that accounts for the observation by postulating individual selection alone. One might expect that if both a group selection and an individual selection hypothesis are consistent with the observations, that one should remain agnostic about which is true. But this is precisely what Williams does not do. He asserts that lower-level selection hypotheses are more parsimonious and therefore ought to be preferred.

An example of this pattern of reasoning is Williams' discussion of dominance/subordination hierarchies in chickens. When chickens are brought together who have not previously lived with each other, there is a good deal of fighting and competitive behavior. However, once a pecking order is established, the rates of egg laying and food consumption increase. Some earlier workers had taken this to be evidence that a dominance hierarchy was a group adaptation; groups competing with other groups fared better when they had a dominance hierachy and that is why present day populations exhibit the characteristic. Williams rejects this explanations on the grounds that it is unparsimonious. For Williams, it is better to suppose that a hierarchy is simply an artifact of the individual struggle for dominance, where the advantages of dominance and subordination are described purely in the currency of the individual organism's survival and reproductive success. Struggle between groups is replaced by the simpler picture of competition within the group.

Williams develops other, perhaps more important, lines of argument in his book.[3] My interest here is just in his parsimony argument. *Adaptation and Natural Selection* was greeted by the biological community as a stunning achievement. This was certainly merited. I do not know whether the parsimony argument was taken to have central importance in Williams' argument, although I suspect that it was. In any event, I do not know of a single case in which biologists at the time expressed scepticism about parsimony. I do not think that biologists asked the inevitable question: why should the greater parsimony of a hypothesis be a ground for thinking that it is more likely to be true? My suspicion is that this question failed to surface because Williams' critique of group selection was thought to be so devastating. Group selection, the idea of adaptations being for the good of the species, and kindred notions were generally regarded as little better than Lamarckism. They were deposited on the rubbish heap of history and the details of the arguments were not examined one by one.

In what follows, I want to explore the logic of Williams' parsimony argument in more detail. I'll suggest that it is in fact a rather weak reason for thinking of the individual organism as the unit of selection. Along the way, several lessons about the role of parsimony in scientific debate may become clearer.

## 3. Units of selection

Perhaps the central conceptual insight in Williams' book is the idea that one should not confuse adaptations with fortuitous benefits. An adaptation is common in a group because there was selection for it. However, once selection drives a trait to fixation, the trait then may provide novel benefits. Applied to the idea of group adaptation, William's idea comes down to this: a trait may benefit the group in which it occurs without its being present *because* it is group-beneficial. If so, it is not a group adaptation and is not to be explained by group selection.

The example described in the previous section – of dominance hierarchies in chickens – illustrates this distinction. Another, purely hypothetical, example that Williams gives is also worth mentioning. If predators catch slow deer more readily than fast ones, there will be selection for fleetness. As a result, the average level of speed found in the group may improve. A fast herd of deer, we may assume, is less likely to go extinct than a slow one. But this group benefit is not the result of group selection; by hypothesis, the only force involved was individual selection for running fast.

The frequency independent fitness functions shown in Figure 1a illustrate this idea. Note that traits $S$ and $A$ have constant fitness values; the frequency of a trait in a group has no impact on how fit, on average, an individual with either trait is. Since $S$ is fitter than $A$ at every frequency, $S$ will go to fixation once it is introduced into a population of $A$ individuals.[4] Notice that $\bar{w}$, the average fitness of individuals in the group, increases in the process. In a sense, the group has become fitter, but not because of group selection. In Williams' example, a fast herd is fitter than a slow one, but this is an artifact of the fact that fast individuals are fitter than slow ones.

Although the frequency independent fitness functions shown in Figure 1a are perhaps very simple, it is well known, both in theory and in practice, that fitnesses can be frequency dependent. In the units of selection controversy, the most commonly discussed example of this is the one shown in Figure 1b. This illustrates the relationship of so-called 'selfishness' ($S$) and 'altruism' ($A$). Note that no matter what the composition of the group is, a selfish individual will be fitter on average than an altruistic one. However, a populatin of altruists will have a higher value of $\bar{w}$ than a population of selfish individuals.

In this case, how fit an individual is depends on two things. First, there is the issue of whether the individual is altruistic or selfish; second, there is the question of the percentage of altruists found within the group the individual inhabits. There is selection for being selfish, but there is also selection for living with altruists. Whether selfishness will reach fixation depends on factors not represented in the fitness

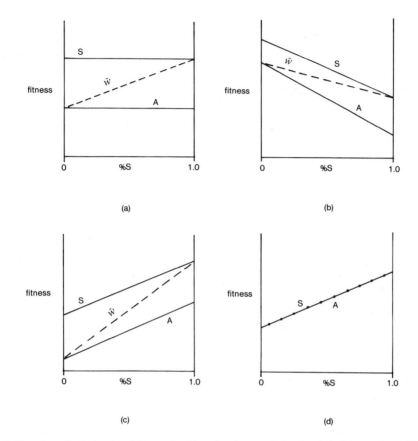

*Fig. 1.* Four hypothetical pairs of fitness functions for characteristics $S$ and A that represent different possible relationships between individual and group selection. In (a), frequency independent individual selection favors $S$, which then goes to fixation within the group. Note that $\bar{w}$, the average fitness of organisms in the group, increases in the process. In (b), group and individual selection oppose each other. Given an ensemble of groups with different local frequencies of $S$, $S$ individuals outcompete A individuals within each group, but groups in which A is common do better than groups in which A is rare. In (c), group and individual selection both act to increase the frequency of $S$. Within each group, $S$ individuals do better than A individuals, and groups in which $S$ is common do better than groups in which $S$ is rare. In (d), there is no variation in fitness within any group; yet, groups in which $S$ is common do better than groups in which $S$ is rare. In this case, there is group selection without individual selection.

function. Group selection will allow altruism to evolve and be maintained if the slope of the fitness functions is steep, if like lives with like, if groups frequently go extinct, and if migrants from altruistic groups frequently found new colonies.

A group selection hypothesis constructed around the fitness function shown in figure 1b will, I guess, be more complicated than the simple scenario suggested by Figure 1a. In Figure 1a, we have frequency independent *individual* selection. If group selection is interpreted as in Figure 1b, we have two mutually opposing selection processes – individual selection favoring selfishness and group selection favoring altruism.

When Williams thought about group selection, he thought about the situation in which group selection *opposes* individual selection. The issue of group selection for him boiled down to the problem of altruism. Williams invoked parsimony to justify his claim that we should postulate group adaptations only if we need to do so. Group selection is reasonable, he thought, only if the characteristic found in nature would be *counter*predicted by the simpler hypothesis of individual selection.

One may perhaps want to accept this suggestion that Figure 1a is preferable to Figure 1b. Or perhaps one may demand specific ecological data for thinking that the fitness function is one way rather than another. Be that as it may, it is important to notice that there is more to the issue of group selection than Figure 1b. We need to consider cases in which group selection does not oppose individual selection.

Figure 1c shows how group and individual selection can both increase the frequency of a characteristic. If the system we were discussing consisted of a single group, group selection would be impossible, since selection requires variation. In this case, $S$ would go to fixation. If, on the other hand, there were various groups that differed in their local frequencies of the traits in question, $S$ would still go to fixation. The cause, in this case, would be two-fold. Regardless of the kind of group an individual is in, the individual will be better off being $S$ than being $A$. And regardless of the individual's phenotype, the individual will be better off when $S$ is common in the group than when $S$ is rare. Here group and individual selection favor the same characteristic.

The final possibility is shown in Figure 1d. David Sloan Wilson has discussed it at length, calling it the 'neutral pathway'.[5] Within a group, $S$ and $A$ have the same fitness values. yet, an individual's fitness is enhanced by belonging to a group in which $S$ is common. If this function applied to a system in which there were a single group, random drift would govern the frequency of the traits, with selection playing no role. If, on the other hand, the system were an ensemble of groups differing in their local frequencies of $S$ and $A$, we would have group selection without individual selection.

Let me make some comments on the relative simplicity of these four fitness functions. Perhaps one thinks that Figure 1a is simpler than Figure 1b, because postulating a single process is simpler than postulating two. If so, one also should think that figure 1d is simpler than both Figures 1b and Figure 1c. Figures 1a and 1d are each quite simple, according to this way of keeping score, in that each posits only a single selection process. Hence, it isn't that individual selection is simpler than group selection, so much as that individual selection by itself (or group selection by *itself*) is simpler than the compound of individual plus group selection. The idea that one process is simpler than two does not generate Williams' judgment that a lower-level selection hypothesis is simpler than a higher-level one.

Of these four figures, three involve frequency dependent fitness values. These three figures all may describe cases in which group selection is at work.[6] Biologists who discuss the idea of frequency dependence usually are taught that Figure 1a is the simplest case. More specifically, they are taught that Figure 1a is usually a *simplification*. Figure 1a is nice, if one wants to prove Fisher's fundamental theorem, for example; but it would be naive to simply assume that nature is always so straightforward.

My final comment concerns the bearing of Figure 1c on a standard way in which parsimony is described. Williams held that one should postulate group selection only if that process is *necessary* to explain the data. Although there may be some general point to this injunction, it leads to a problem in this specific case. The fixation of trait *S* is *overdetermined* in Figure 1c. Group selection isn't *necessary* here; even if there were only a single group, *S* would still go to fixation. Yet fitness functions like Figure 1c *can* describe cases in which group selection occurs. A full investigation of the importance of group selection in nature would attend to Figure 1c as well as to Figures 1b and 1d. By restricting our attention to cases in which group selection opposes individual selection, we bias our inquiry.

There may, of course, be a methodological point to doing this. Figure 1b makes an inference of group selection far more clear-cut, since in this case individual selection *counter*predicts the evolution and maintenance of altruism. My point is that although Figure 1b may make matters easier, an exclusive attention to it will fail to be complete.

The discussion of group selection has, since Williams' book, moved beyond the issue of parsimony. Even if we could convince ourselves that Figure 1a is the simplest of the four fitness functions, no biologist ought to think that this is much of a reason for discounting group selection. To decide this question, biologists examine contingent properties of population structure. They try to get empirical evidence as to which of these four functions is exemplified in nature; they also try to get empirical evidence concerning the other parameters, like extinction and colonization, that play a role. This is entirely as it should be. In the absence of such data, I think we ought to be agnostic with respect to the question of group selection. Even if parsimony were a reason here, it would not be a very compelling one.

### 4. Concluding remarks

The use of parsimony in the units of selection problem has been scientifically influential. This, by itself, is a reason for subjecting it to careful philosophical scrutiny. But philosophy of science quite naturally aims at extracting more general lessons from particular examples. In conclusion, I'll say a few words in this vein.

I already have mentioned one such consequence. The principle of parsimony is often thought to say that one should not postulate the existence of something unless such postulation is *necessary* to explain one's observations. This idea obviously played a role in Williams' investigation of the units of selection problem; it led him to focus on characteristics that allegedly could not evolve unless there was group selection. In this way, the group selection problem was reduced to the problem of altruism. I have discussed how this gave too narrow a focus to the problem of group selection. Indeed, it is entirely possible that group selection was one of the forces influencing the evolution of a character, even though it is *false* that group selection was necessary for the character to evolve. The formulation of Ockham's razor in terms of the idea of *explanatory necessity* is at best a rough first approximation.

Another consequence suggested by the units of selection problem concerns the *weight* accorded to parsimony as opposed to other considerations. For biologists interested in discovering *why* group selection is common or rare, rather than just *whether* it is so, parsimony must be at best a very incomplete perspective. Even if group selection hypotheses were unparsimonious, this would not explain *why* group selection rarely occurs.[7] This latter question, which has been the main preoccupation of biologists interested in the units of selection *after* the publication of Williams' book, has led to the construction of models in which the possibility and import of group selection are a function of contingent properties of population structure.[8] Clearly, the importance accorded parsimony depends on one's research agenda.

Even within the context of deciding whether group selection has occurred in a particular context, parsimony seems to be a temporary expedient rather than a sufficient reason. Even if we were to follow Williams and grant that parsimony can be a reason against some group selection hypothesis, how much better a reason would be provided by independent empirical confirmation of the details of one selection scenario as opposed to the other. And not only would this further, empirical, approach *improve* the evidential warrant of some conclusion about group selection; it is perhaps not too much an overstatement to say that without this further kind of consideration, the conclusion licensed by parsimony alone would be rather uncompelling.

All this is not to suggest that parsimony is peripheral to the enterprise of science. Perhaps underlying the entire edifice of knowledge is a methodological canon of this sort. However, what we see in the units of selection controversy is an example of how parsimony functions *in* science, not in some hypothetical substratum upon which science is said to rest. Work in the theory of nondeductive inference may illuminate the latter topic and allow us to see how general and *a priori* parsimony constraints, if such there be, are related to the specific appeals to parsimony that from time to time surface in empirical research.

**Notes**

1. Quine (1953).
2. Williams (1966); Dawkins (1976) also gives parsimony considerable weight.

3. I discuss these in Sober (1984c).
4. I assume that like always produces like, to simplify things.
5. Wilson (1980).
6. I do not claim that frequency dependent selection suffices for group selection, for reasons given in Sober (1984).
7. A point I develop in Sober (1984).
8. Some of these are reviewed in Wade (1978).


## Bibliography

Dawkins, R. (1976), *The Selfish Gene*. (Oxford: Oxford University Press).

Quine, W. (1953), 'Two Dogmas of Empiricism'. In *From a Logical Point of View*. New York: Harper Torchbooks.

Rosenkrantz, R. (1976), 'Simplicity'. In *Foundations and Philosophy of Statistical Inference,* W. Harper and C. Hooker, eds. (Boston: Reidel).

Sober, E. (1975), *Simplicity*. (Oxford: Oxford University Press).

Sober, E. (1984), *The Nature of Selection: Evolutionary Theory in Philosophical Focus*. (Cambridge: Bradford/MIT Press).

Wade, M. (1978), 'A Critical Review of the Models of Group Selection'. *Quarterly Review of Biology* 53: 101–114.

Williams, G. (1966), *Adaptation and Natural Selection*. (Princeton: Princeton University Press).

Wilson, D. (1980), *The Natural Selection of Populations and Communities*. (Menlo Park, California: Benjamin/Cummings).

## SCIENCE AND PHILOSOPHY

Nersessian NJ: Faraday to Einstein – Constructing Meaning in Scientific Theories. 1984. ISBN 90-247-2997-1.

Bechtel PW (ed): Integrating Scientific Disciplines. 1986. ISBN 90-247-3242-5.

Nersessian NJ (ed): The Process of Science. 1987. ISBN 90-247-3425-8.